Saltmarshes

Morphodynamics, Conservation
and Engineering Significance

Saltmarshes

Morphodynamics, Conservation and Engineering Significance

EDITED BY

J.R.L. ALLEN AND K. PYE

Postgraduate Research Institute for Sedimentology, University of Reading

Published by the Press Syndicate of the University of Cambridge,
The Pitt Building, Trumpington Street, Cambridge, Cambridge CB2 1RP
40 West 20th Street, New York, NY 10011-4211, USA
10 Stamford Road, Oakleigh, Victoria 3166, Australia

First published 1992

Printed in Great Britain at the University Press, Cambridge

A catalogue record of this book is available from the British Library

Library of Congress cataloguing in publication data available

ISBN 0 521 41841 0 hardback

GE

Contents

Contributors

J.R.L. ALLEN, Postgraduate Research Institute for Sedimentology, University of Reading, P.O. Box 227, Whiteknights, Reading RG6 2AB.

A.H. BRAMPTON, Hydraulics Research Ltd, Wallingford, Oxon OX10 8BA.

J.P. DOODY, UK Joint Nature Conservation Committee, Monkstone Hill, 3rd floor, City Road, Peterborough PE1 1JY.

A.J. GRAY, Institute of Terrestrial Ecology, Furzebrook Research Station, Wareham, Dorset BH20 5AS.

J.S. PETHICK, Institute of Estuarine and Coastal Studies, University of Hull, Hull HU6 7RX.

K. PYE, Postgraduate Research Institute for Sedimentology, University of Reading, P.O. Box 227, Whiteknights, Reading RG6 2AB.

M.J. TOOLEY, Environmental Research Centre, Department of Geography, University of Durham, Science Laboratories, Durham DH1 3LE.

Preface

Coastal saltmarshes are areas of land covered principally by halophytic vegetation which are regularly flooded by the sea. They occur in many temperate and high-latitude estuaries and on sections of open coast which are protected from extreme wave action by wide intertidal flats and barrier complexes. Extensive areas of reclaimed coastal marsh, now largely fresh, are also found in many countries, including Great Britain.

Attitudes towards saltmarshes have traditionally been diverse. They have long been valued by ecologists, ornithologists and conservationists as important wildlife habitats, and as field laboratories for scientific investigation. Marshes and associated mudflats provide feeding grounds and nesting sites for a wide range of wading birds, and act as important migration stop-over sites. Academic geomorphologists, sedimentologists and geochemists have also valued saltmarshes as dynamic natural environments in which the interaction of natural physical, chemical and biological processes can be observed, monitored and demonstrated for teaching purposes. In a commercial sense, saltmarshes have been valued by farmers as sites for seasonal grazing and wild-fowling, or as potential sites for reclamation and arable cultivation. On the other hand, industrialists and local planning authorities have often viewed saltmarshes as areas of low economic value suitable only for waste-dumping and industrial development. The majority of the general public have also preferred to choose sandy and rocky coasts for recreational and residential development, and thought it appropriate to choose saltmarshes as sites for petrochemical complexes and dock developments. The attitude of engineers to saltmarshes has traditionally reflected the objectives of their clients, which have principally been to enclose, drain and reclaim marshes, and to limit their growth as undesirable influences on navigation and beach quality. However, in recent years the potential value of saltmarshes in coast protection and flood defence has been increasingly recognized. Concern about the effects of accelerated sea-level rise and a possible increase in storminess which have been predicted to occur during the next century, coupled with a decline in the value of agricultural land and the increasingly vociferous arguments being forwarded by environmental pressure groups, has caused the authorities responsible for flood defence and coast protection to review their current strategies. Options of land abandonment and 'set-back' of sea walls on a large scale are now being actively considered, and there is renewed interest in the development of new 'soft-engineering' methods for saltmarsh creation and restoration. This growth of interest has revealed that many of the basic physical and biological processes governing the formation and dynamics of saltmarshes remain poorly understood and require further study.

It was against this background that a one-day workshop on the 'Morphodynamics, Conservation and Engineering Significance of Saltmarshes' was held at the Postgraduate Research Institute for Sedimentology, University of Reading, on 24 April 1991. Seven leading authorities in British coastal saltmarsh research were invited to summarise the current status of knowledge in their specialist subject areas prior to a discussion of outstanding problems and requirements for further work. The meeting was attended by over fifty participants drawn from a wide range of university departments, Government research establishments, the National Rivers Authority, and engineering consultancy companies. This book, which is based on the formal contributions to the workshop, has been produced with the intention of making available the information presented to a wider international audience.

J.R.L. Allen K. Pye Reading May 1991

Coastal saltmarshes: their nature and importance

J.R.L. ALLEN AND K. PYE

Introduction

Coastal saltmarshes are environments high in the intertidal zone where a generally muddy substrate supports varied and normally dense stands of halophytic plants. These environments grade seawards into mudflats or sandflats, to which they are genetically related, and from which they are often separated by either a ramp or cliff, and may grade upwards and landwards into freshwater marshes and coastal woodland communities. Saltmarshes are widely developed on low-energy coasts in temperate and high latitudes, but in the tropics and sub-tropics they are replaced by mangrove communities. The approximate latitudinal limit for the growth of mangrove species is determined by a mean minimum temperature in the coldest month of 10°C (Chapman, 1977).

Saltmarshes have long been viewed by scientists as intrinsically interesting environments on account of their variability and the rapidity with which physical, chemical and biological processes operate. A number of reviews have been published during the course of the past century (Carey and Oliver, 1918; Chapman, 1960; Ranwell, 1972; Beeftink, 1977; Long and Mason, 1983; Frey and Basan, 1985; Adam, 1990). However, much of the published work has been ecological or geochemical in character, and physical processes have been relatively neglected.

The occurrence of saltmarshes is to a large extent controlled by coastal physiography, since under most circumstances mud can accumulate only in relatively low-energy environments where wave action is limited. Consequently mudflats and marshes are usually found in sheltered embayments and estuaries, and in the lee of barrier islands and spits. An exception occurs in areas where very large amounts of fine sediment are supplied to the coastal zone by major rivers, resulting in the formation of a wide and shallow nearshore zone which absorbs much of the incoming wave energy. In such circumstances, such as occur in the Mississippi Delta and northwest

Fig. 1.1. Generalised distribution of active saltmarshes around the British coast (modified after Burd, 1989).

of the mouth of the Amazon, muddy sediments can accumulate on parts of the open coast and may evolve into saltmarshes or mangrove forests.

The total area covered by active saltmarsh in Great Britain amounts to approximately 44,370 ha, being concentrated largely in eastern and southeastern England, northwest England and the area of the Bristol Channel (Burd, 1989; Fig. 1.1). Active saltmarshes are also locally important in Northern Ireland. Major areas of reclaimed marsh are found in East Anglia, Kent, Somerset, and northwest Lancashire (Gray, 1977). The active marshes can be classified into five main types on the basis of their physical setting: (1) open coast marshes, (2) back-barrier marshes, (3) estuarine fringing marshes, (4) embayment marshes, and (5) loch or fjord-head marshes. True open-coast marshes are poorly developed in Britain on account of the relatively high wave energy experienced along most of the coast, examples being found mainly in Essex along the Dengie Peninsula and on Foulness Island. Open-coast back-barrier marshes are well developed on the shores of north Norfolk (Pye, Chapter 8, this volume) and in south Lincolnshire, but are also otherwise poorly developed in the UK. Estuarine fringing marshes occur in virtually every estuary, including the Severn (Allen, Chapter 7, this volume), Dee, Mersey, Ribble and Solway on the west coast, and the Medway, Thames, Crouch, Blackwater, Humber and Tay on the east coast. Embayment marshes are found in relatively large, shallow coastal embayments which often have a restricted entrance and receive a relatively limited freshwater input. In some instances the entrance to the embayment is partly protected by a sand or shingle barrier. Several examples are found on the south coast of England, including Portsmouth Harbour, Langstone Harbour, Chichester Harbour, Pagham Harbour and Poole Harbour. An east-coast example is provided by Hamford Water, while The Wash and Morecambe Bay marshes can also be regarded as variants of embayment-type marshes. Loch- or fjord-head marshes are typically restricted in size and occur mainly on the predominatly rocky coasts found in northwest Scotland.

The location, character and dynamic behaviour of saltmarshes is governed essentially by four physical factors: sediment supply, tidal regime, wind-wave climate, and the movement of relative sea level. To these may be added the variable but secondary role played by marsh vegetation in acting both as a source and as a trapper and binder of sediment. Colonisation of sand or mudflats by vegetation can only begin once the level of the surface has been raised to a sufficiently high level in the tidal frame by physical sedimentation processes. Once vegetation is established, the rate of sedimentation frequently increases as more of the incoming sediment is intercepted and trapped by the greater surface roughness (Stumpf, 1983; Stevenson *et al.*, 1988), resuspension of deposited material is reduced for the same reason, and organic matter is added to the marsh surface.

Sediment supply

The immediate sources of the sediment found beneath saltmarshes are the tidal waters (which provide mainly mineral matter) and the marsh plants themselves (which supply organic matter). The composition and grain size of the mineral matter varies from sandy silt to clayey silt according to marsh location and marsh height, but is normally referred to as 'mud'. A predominance of mud in the supply leads to the formation of a *minerogenic* marsh, whereas a predominance of organic matter supply (litter, root biomass) leads to the formation of an *organogenic* marsh. The saltmarshes actively forming in Great Britain today are mostly of the minerogenic type.

Relatively little is known about the sources and budgets of fine sediment in British saltmarshes, or about the manner in which these have varied over time during the Quaternary. Potential sources of minerogenic sediment include river catchments, estuarine and coastal cliffs, and offshore mud deposits. Since mud is easily transported in suspension in tidal waters, it may travel considerable distances from its source, and be mixed with material derived from other sources, before arriving at its ultimate site of deposition. The major rivers which drain into the North Sea and Irish Sea at the present day supply relatively small amounts of sediment (McCave, 1987; Eisma and Kalf, 1987; Kirby, 1987), and much of the mud suspended in tidal waters appears to be derived from erosion of unconsolidated Pleistocene glacial sediments exposed in coastal cliff exposures. During the early Flandrian period, when sea level rose rapidly, wave and tidal current reworking of Pleistocene sediments on the floor of the North Sea and Irish Sea probably provided a major source of sediment which was reworked landwards and deposited in protected estuaries and embayments. Most British estuaries have acted as long-term sediment sinks throughout the Flandrian period and many have experienced a marked reduction in estuarine capacity, due in part to natural sedimentation processes but enhanced by human activities which have included reclamation and dredging (Kestner, 1979; O'Connor, 1987).

The margins of certain estuaries, for example, the Severn, expose substantial thicknesses of postglacial silts which, representing earlier marshes and mudflats, are now undergoing vigorous erosion. Part of the sediment derived from these sources finds its way onto younger marshes further up the estuary (e.g. Allen, 1990).

The availability of suspended fine sediment, and the maximum productivity of marsh plants, together strongly affect the character of saltmarshes and their ability to respond to changing environmental conditions.

Tidal regime

Mud accumulates high in the intertidal zone of tide-dominated coasts because it is only at stages close to high water level that tidal current velocities are sufficiently low to allow fine suspended particles to settle out and remain undisturbed. Since much of the coastline of Britain is either mesotidal or macrotidal, the vertical range of saltmarshes within the high intertidal zone is typically 1–4 m. In most areas, marsh vegetation is limited to the zone between mid neap tide level and high water spring tide level. Most of the British coast experiences a simple semi-diurnal tidal regime, although on parts of the south coast the tidal regime is more complex.

The combination of a macro- or mesotidal regime and comparative shallowness means that most British barrier coasts, embayments and estuaries are dominated by a flood-tide regime which favours the landward movement of sediment from the offshore zone and which encourages the retention of river-borne sediment close inshore.

Due to the fact that saltmarsh and mudflat sediments are exposed to the atmosphere at low tide, their geotechnical characteristics differ from those of fine sediments which accumulate wholly subaqueously. Drying between tides and during the summer season of comparatively low tides gives intertidal muddy sediments a head start on the path to consolidation before any significant load has been experienced. Additionally, drying creates fractures which frequently influence the pattern of erosion during subsequent high tides. Such fractures can play a major role in determining the nature and rate of marsh cliff retreat (Allen, 1989).

Wind and wave climate

Although tide-dominated, the barrier coasts, embayments and estuaries of Great Britain and Northern Ireland are not devoid of wave influences. The effectiveness of waves in these environments depends on wind strength and directional variability, fetch (distance of open water over which waves can be generated), and on the frequency distribution of water stages. Powerful waves can quarry large blocks of sediment and cause undercutting and collapse of saltmarsh cliffs. They may also destroy the integrity of the surface vegetation, particularly near the seaward marsh edge, leading to widespread scouring of the underlying mud surface and promoting a generally unstable local environment. Waves and wave-induced bottom currents may prevent the settlement of mud at high tide, and may resuspend material deposited by earlier tides. Storm waves may also influence saltmarshes through the construction or destruction of shingle barriers (Fig. 1.2).

Although wave height tends to increase linearly with wind speed, wave energy varies as the square of wave height; consequently relatively modest fluctuations in local wind-wave climate may bring about substantial changes in the high intertidal zone.

Instrumental and proxy records demonstrate that climate in the British Isles has experienced considerable temporal variability on a variety of timescales during the last few thousand years (Lamb, 1982). Saltmarshes and the mudflats from which they evolve may well prove to afford a sensitive historical record of the local effects of these fluctuations (Allen, Chapter 7, this volume).

Movements of relative sea level

The behaviour of relative sea level can have a major influence on the medium and long-term evolution of saltmarshes (Reed, 1990). The causes of relative sea-level movements can be divided into three main groups: (1) eustatic factors, which are essentially global in extent, including changes in ice volume, thermal expansion/contraction of the oceans, and changes in ocean volume due to sea floor spreading and hydrostatic loading (Warrick and Oerlemans, 1990; Wigley and Raper, 1991); (2) regional factors, including subsidence/uplift due to crustal movements and sediment compaction/dewatering, geoid changes and tidal variations; and (3) local factors, including changes

Fig. 1.2. Storm-generated shingle washover fan, composed largely of shell debris, transgressing over marsh vegetation, north side of the Blackwater Estuary, Essex.

in coastal, estuarine and shelf morphology, which may affect tidal regime (Woodworth *et al.*, 1991), and changes in barometric pressure and wind field, which in turn have an influence on mean water levels (Woodworth, 1987, 1990; Pugh, 1990).

Relative sea-level movements in the British Isles over at least the past few centuries appear to have been dominated by regional neotectonic factors, but the eustatic contribution due to thermal expansion of the oceans and decreasing ice volume may become more significant in the next century (Warrick and Oerlemans, 1990; Warrick and Wigley, 1991) The present net sea-level trend remains downward in northwest Britain, where the land is still rising isostatically following removal of the former ice load approximately 10,000 years ago, but in the southern part of the country the trend is upward, albeit apparently locally variable (Fig. 1.3; Woodworth, 1987, 1990; Tooley, Chapter 2, this volume). This rise in relative sea level can in part be attributed to long-term subsidence of the southern North Sea Basin, but the relative contribution of other factors remains a matter of controversy. A zone of neutral crustal movement (and, effectively, sea-level movement) has been identified running across the country

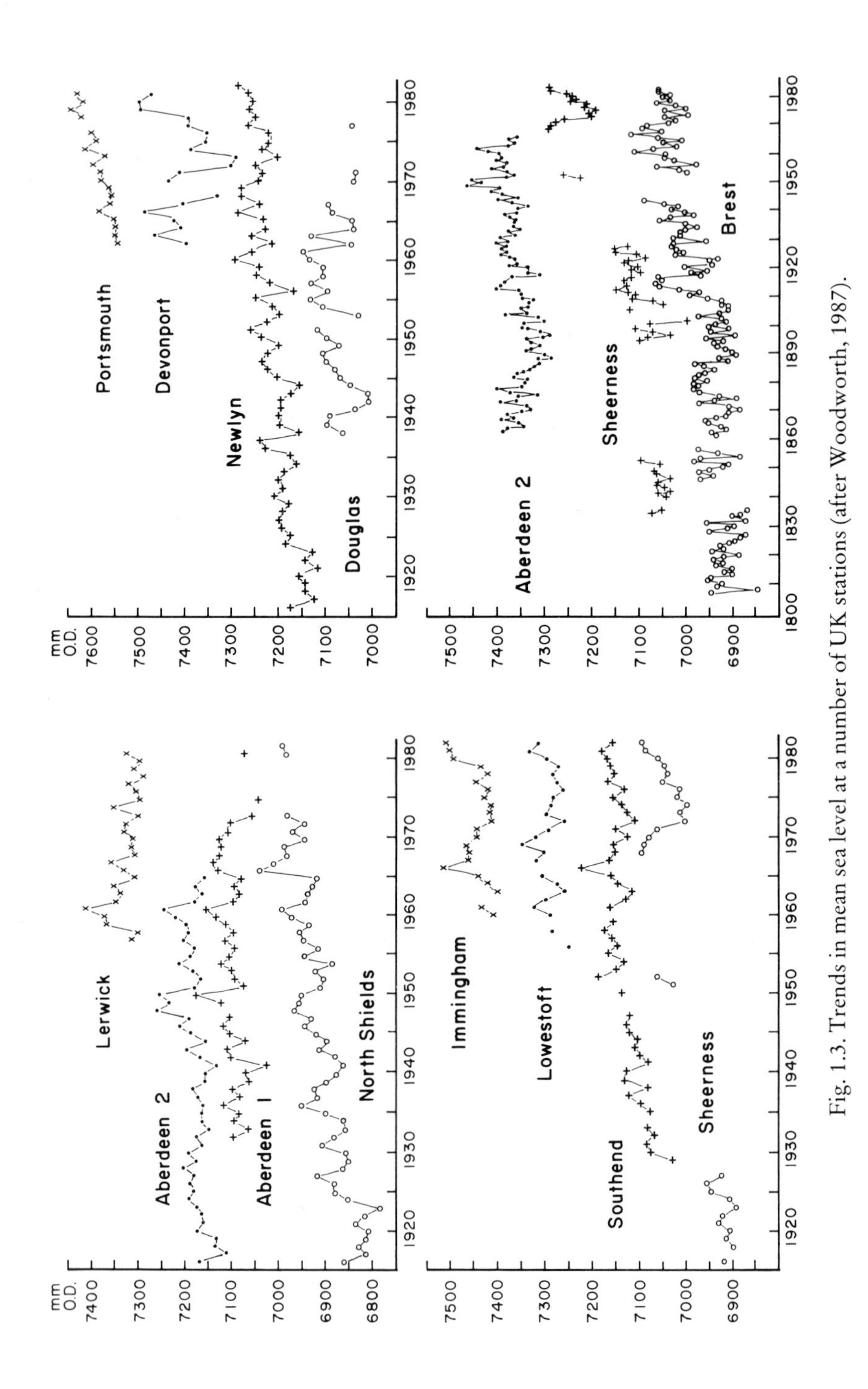

Fig. 1.3. Trends in mean sea level at a number of UK stations (after Woodworth, 1987).

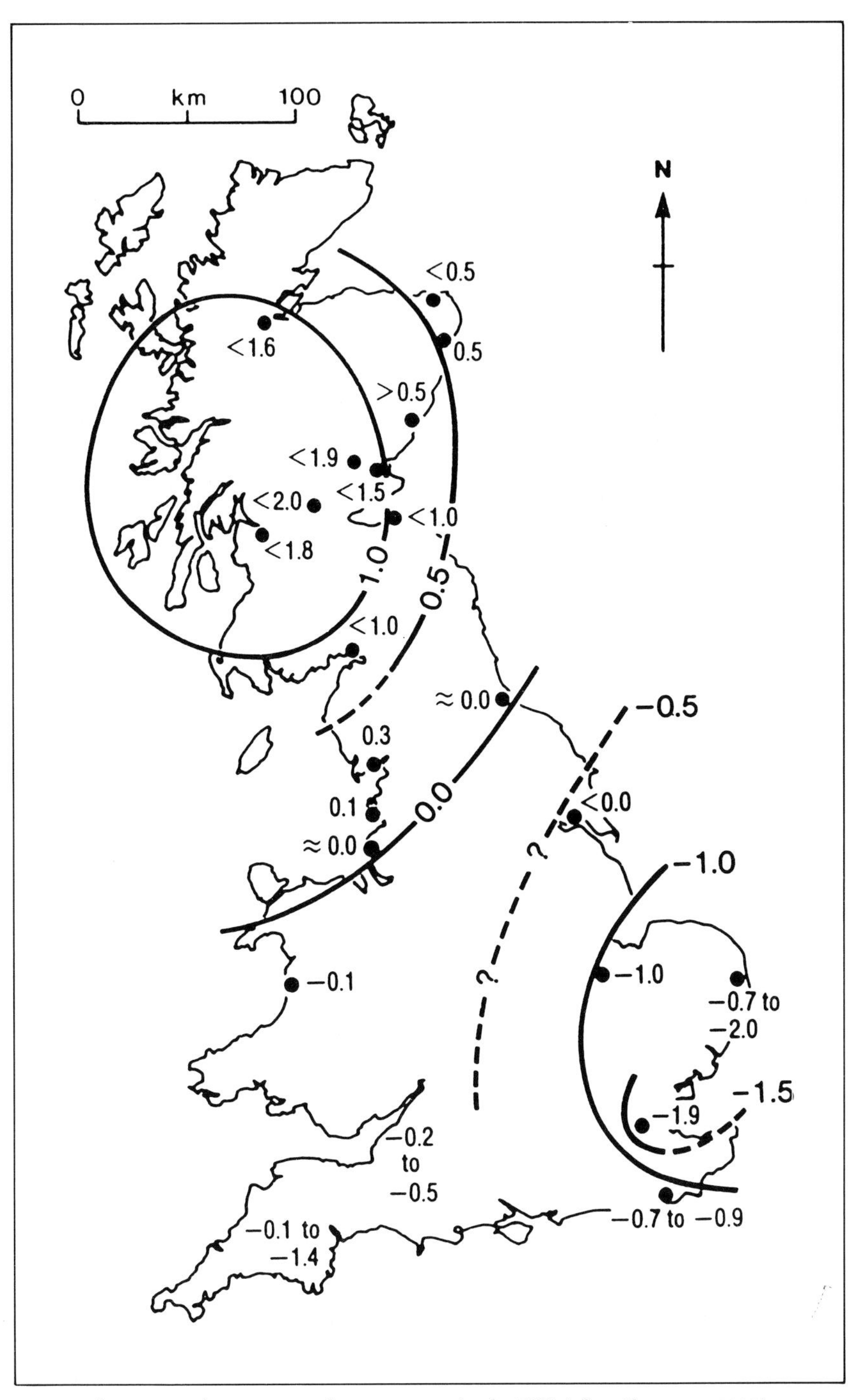

Fig. 1.4. Holocene crustal movements in the UK (after Shennan, 1989).

Fig. 1.5. Wave-eroded marsh cliff, Stiffkey, Norfolk.

roughly from the Mersey to The Tees (Fig. 1.4; Shennan, 1989). The increase in the rate of global sea-level rise forecast during the next century will serve to shift the zone of neutral sea level change northwards and will enhance the upward sea-level trend in southern England, where many major saltmarsh complexes are located.

Whereas wind-wave climate has most influence on the horizontal extent of mudflats and saltmarshes, sea-level, combined with tidal range (which may be expected to vary with relative sea level), mostly affects their vertical growth. Under normal circumstances, saltmarshes do not grow higher than the level of the highest astronomical tide, since the frequency and duration of tidal flooding controls the introduction of minerogenic sediment. In northern Britain, where relative sea level is falling, saltmarshes have the potential, once they have reached a high enough position in the tidal frame, to evolve into brackish or freshwater organogenic marshes. and then into coastal scrub or woodland. By contrast, in southern Britain, saltmarshes have the task of maintaining a combined rate of vertical minerogenic and organogenic accretion which can keep pace with the rate of sea level rise. The physical and ecological response of saltmarshes to the forecast increase in rate of sea-level rise in the next century is of widespread concern (Boorman *et al.*, 1989; Department of the Environment, 1991). Many saltmarshes in Essex, Kent and parts of the South Coast are already experiencing serious erosion (Kirby, 1985; Harmsworth and Long, 1986), and there is concern that the marshes may deteriorate further and, in some areas, eventually disappear.

Saltmarsh erosion in these areas currently takes several forms; (1) erosion of the marsh edge, forming either a steep cliff (Fig. 1.5) or ramp on which erosional spur and groove topography is well developed (Fig. 1.6; Pethick, Chapter 3, this volume); (2) enlargement of pans and creeks within the marsh by bank collapse and headwall retreat, leading to coalescence and appearance of extensive areas of bare mudflat with residual vegetated hummocks (Fig. 1.7); (3) widespread deterioration of marsh vegetation, leading to generalised scour and surface lowering. All of these modes of marsh erosion have a profound effect on the vegetation ecology and conservation value of the saltmarshes (Gray, Chapter 4; Doody, Chapter 5, this volume). A reduction in the width and/or height of the saltmarshes also reduces their effectiveness in damping waves and increases the risk of overtopping or breaching of sea defence structures on their landward side (Brampton, Chapter 6, this volume). As

Fig. 1.6. Erosional spurs and grooves on the seaward margin of the Dengie marshes, Essex.

yet, however, it is by no means established with certainty that increased marsh erosion in southeastern England is associated with changing sea level; the role of changes in inshore wave energy, sediment supply and the effects of human activities including dredging and channel modification have not yet been fully evaluated.

Outstanding questions about saltmarshes and requirements for future research

Although saltmarshes represent an attractively complex and relatively accessible environment, and have consequently been studied from several standpoints over many decades, many of the fundamental questions concerning them (and the intertidal flats from which they develop) remain incompletely answered. In the context of British saltmarshes these questions include:

(1) What are the sources and amounts of fine sediment currently supplied to particular saltmarsh complexes?
(2) Can geochemical, mineralogical and physical characteristics be used to identify sediment transport paths and quantify sediment budgets?

Fig. 1.7. Residual mud mounds formed by internal marsh erosion, Tollesbury marshes, Essex.

(3) What are the basic depositional and erosional mechanisms, as governed by the tide, waves and plants, involved in the formation of saltmarshes and mudflats?

(4) What are the current rates of vertical and lateral accretion/erosion on British saltmarshes?

(5) What is the degree of geomorphological and sedimentological variability of British saltmarshes and mudflats, and what is its explanation?

(6) What is the relative importance of factors which control the vigour of saltmarsh vegetation, and why are certain species prone to periodic die-back? Can this be managed and controlled?

(7) In what ways, and to what degrees, have saltmarshes and mudflats responded to historical changes in such forcing factors as tidal regime, wind-wave climate and sea-level movement, and what has been the importance of essentially local factors, such as shifting channels and shoals?

(8) Can models be developed which will predict the likely response of existing marshes to changes in relative sea level, wind-wave climate, and sediment supply over the next century,

Fig. 1.8. Marsh restoration works, Deal Hall, Dengie Peninsula, Essex.

and can these models be developed sufficiently quickly that they can be incorporated into cost-benefit analyses relating to future coastal management strategies?

(9) Can measured short-term accretion rates on marshes be used as indicators of the rate of sea-level rise? Do we understand enough about post-depositional modification and consolidation of marsh sediments, and about spatial and temporal variations in accretion rates, to identify a meaningful sea-level 'signal' against background 'noise'?

(10) Can monitoring of changes in marsh vegetation give an early warning of stress to the system induced by sea-level rise or other environmental changes (Vanderzee, 1988)?

(11) What are the implications of increased marsh erosion in polluted estuarine systems? Will this result in a pulse of contaminants being released into the water column? If so, what are the likely effects?

(12) What effects will accelerated sea-level rise and possible changes in wind/wave climate have on marsh vegetation communities and their associated fauna? Will there be an increase in the area covered by low and mid-marsh communities at the expense of

Fig. 1.9. Experimental planting of *Spartina* on artificial mud mounds, Dengie Peninsula, Essex.

high-marsh communities, or will low-marsh and mudflat environments become less extensive in response to a steepening of the nearshore profile? How will these changes affect migratory birds and other species?

(13) What methods are most appropriate in the British context for the artificial creation and rehabilitation of eroding saltmarshes? Are the 'polder' technologies which have been developed in North Germany, The Netherlands and Denmark likely to prove successful in the UK context (Figs. 1.8, 1.9)? Is 'marsh nourishment', involving pumping of sediment onto marshes from adjacent mudflats or from imported dredge spoil (Delaune *et al.*, 1990), a viable alternative? How successful are artificial roughness elements and wave breaks (Seneca *et al.*, 1975) in accelerating the accretion rate on mudflats and saltmarshes? Is mudflat 'ploughing' a worthwhile and effective way of increasing the suspended sediment concentration and marsh accretion rates? What are the best methods of establishing vegetation on unvegetated mudflats under different conditions? What methods should be used to deal with eroding saltmarsh cliffs? Are the engineering methods devised for

application in fluvial systems (Coppin and Richards, 1990) appropriate for intertidal systems?

(14) What will be the physical and ecological effects of abandoning or repositioning existing sea walls? Is this a cost effective, environmentally acceptable and politically feasible policy?

A number of recent reports have gone some way towards establishing the basis on which these questions can be addressed (Doody, 1985; Hydraulics Research, 1987, 1988; Boorman *et al.*, 1989; Gray and Benham, 1990; Cannell and Hooper, 1991; Department of Environment, 1991), but much fundamental research remains to be carried out before definitive answers can be given.

Acknowledgements

We are grateful to Rosemary Gillespie for assistance in preparing this and the other written contributions arising from the Saltmarsh workshop. University of Reading PRIS Contribution No. 141.

References

ADAM, P. (1990) *Saltmarsh Ecology*. Cambridge University Press, Cambridge, 461 pp.

ALLEN, J.R.L. (1989) Evolution of salt-marsh cliffs in muddy and sandy systems: a qualitative comparison of British west-coast estuaries. *Earth Surface Processes and Landforms* **14**, 85–92.

ALLEN, J.R.L. (1990) The Severn Estuary in southwest Britain: its retreat under marine transgression, and fine-sediment regime. *Sedimentary Geology* **96**, 13–28.

BEEFTINK, W.G. (1977) Salt marshes. In R.S.K. Barnes (ed.) *The Coastline*. Wiley, London, 93–121.

BOORMAN, L.A., GOSS-CUSTARD, J. & MCGORTY, S. (1989) *Climate Change, Rising Sea Level and the British Coast.* NERC Institute of Terestrial Ecology Research Publication No. 1. HMSO, London, 24 pp.

BURD, F. (1989) *The Saltmarsh Survey of Great Britain. An Inventory of British Saltmarshes.* Research and Survey in Nature Conservation No. 17. Nature Conservancy Council, Peterborough, 180 pp.

CANNELL, M.G.R. & HOOPER, M.D. (1991) *The Greenhouse Effect and Terrestrial Ecosystems of the UK.* NERC Institute of Terrestrial Ecology Research Publication No. 4. HMSO, London, 56 pp.

CAREY, A.E. & OLIVER, F.W. (1918) *Tidal Lands. A Study of Shore Problems.* Blackie, Glasgow.

CHAPMAN, V.J. (1960) *Salt Marshes and Salt Deserts of the World.* Leonard Hill, London.

CHAPMAN, V.J. (ED.) (1977) *Wet Coastal Ecosystems.* Elsevier, Amsterdam.

COPPIN, N.J. & RICHARDS, I.G. (EDS) (1990) *Use of Vegetation in Civil Engineering.* Butterworths, London.

DELAUNE, R.D., PEZESHKI, S.R., PARDUE, J.H., WHITCOMB, J.H. & PATRICK, W.H. (1990) Some influences of sediment addition to a deteriorating salt marsh in the Mississippi River deltaic plain: a pilot study. *Journal of Coastal Research* **6**, 181–188.

DEPARTMENT OF ENVIRONMENT (1991) *The Potential Effects of Climate Change in the United Kingdom.* HMSO, London, 124 pp.

DOODY, J.P. (ED.) (1985) *Spartina anglica in Great Britain.* Focus on Nature Conservation No. 7, Nature Conservancy Council, Peterborough.

EISMA, D. & KALF, J. (1987) Dispersal, concentration and deposition of suspended matter in the North Sea. *Journal of the Geological Society* **144**, 161–178.

FREY, R.W. & BASAN, P.B. (1985) Coastal salt marshes. In R.A. Davis Jr (ed.) *Coastal Sedimentary Environments* 2nd edn. Springer, New York, 225–301.

GRAY, A.J. (1977) Reclaimed land. In R.S.K. Barnes (ed.) *The Coastline.* Wiley, London, 253–270.

GRAY, A.J. & BENHAM, P.E.M. (EDS) (1990) *Spartina anglica – A Research Review.* NERC Institute of Terrestrial Ecology Research Publication No. 2. London, HMSO, 80 pp.

HARMSWORTH, G.C. & LONG, S.P. (1986) An assessment of saltmarsh erosion in Essex, England, with reference to the Dengie Peninsula. *Biological Conservation* **35**, 377–387.

HYDRAULICS RESEARCH LTD (1987) *The Effectiveness of Saltings.* Report No. SR 109. Hydraulics Research Ltd, Wallingford.

HYDRAULICS RESEARCH LTD (1988) *Review of the Use of Saltings in Coastal Defence.* Report No. SR 170. Hydraulics Research Ltd, Wallingford.

KESTNER, F.J.T. (1979) Loose boundary hydraulics and land reclamation. In B. Knights & A.J. Phillips (eds) *Estuarine and Coastal Land Reclamation and Water Storage.* Saxon House, Farnborough, 23–47.

KIRBY, R. (1985) The recent history of the lower Medway saltmarshes. In J.P. Doody (ed.) *Spartina anglica in Great Britain.* Focus in Nature Conservation No. 5, Nature Conservancy Council, Peterborough, 18–20.

KIRBY, R. (1987) Sediment exchanges across the coastal margins of N.W. Europe. *Journal of the Geological Society* **144**, 121–126.

LAMB, H.H. (1982) *Climate, History and the Modern World.* University Paperbacks, London.

LONG, S.P. & MASON, C.F. (1983) *Saltmarsh Ecology*. Blackie, Glasgow.

McCAVE, I.N. (1987) Fine sediment sources and sinks around the East Anglian coast. *Journal of the Geological Society* **144**, 149–152.

O'CONNOR, B.A. (1987) Short and long term changes in estuary capacity. *Journal of the Geological Society* **144**, 187–196.

PUGH, D.T. (1990) Is there a sea-level problem? *Proceedings of the Institution of Civil Engineers* Part 1. **88**, 347–366.

RANWELL, D.S. (1972) *Ecology of Salt Marshes and Sand Dunes.* Chapman and Hall, London.

REED, D. J. (1990) The impact of sea level rise on coastal salt marshes. *Progress in Physical Geography* **14**, 483–481.

SENECA, E.D., WOODHOUSE, W.W. & BROOME S.W. (1975) Salt-water marsh creation. In L.E. Cronin (ed.) *Estuarine Research, Volume 2.* Academic Press, New York, 427–438.

SHENNAN, I. (1989) Holocene crustal movements and sea level changes in Great Britain. *Journal of Quaternary Science* **4**, 77–89.

STEVENSON, J.C., WARD, L.G. & KEARNEY, M.S. (1988) Sediment transport and trapping in marsh systems: implications of tidal flux studies. *Marine Geology* **80**, 37–59.

STUMPF, R.P. (1983) The process of sedimentation on the surface of a salt marsh. *Estuarine Coastal and Shelf Science* **17**, 495–508.

VANDERZEE, M.P. (1988) Changes in saltmarsh vegetation as an early indicator of sea level rise. In G.I. Pearson (ed.) *Greenhouse – Planning for Climate Change.* CSIRO, Melbourne, 147–160.

WARWICK, R.A. & OERLEMANS, H. (1991) Sea level rise. In J.T. Houghton, G.J. Jenkins & J.J. Ephramus, (eds) *Climate Change, The IPCC Scientific Assessment.* Cambridge University Press, Cambridge, 259–281.

WIGLEY, T.M.L. & RAPER, S.C.B. (1992) Future changes in global mean temperature and thermal-expansion-related sea level rise. In R.A. Warrick & T.M.L. Wigley (eds) *Climate and Sea Level Change: Observations, Projections and Implications.* Cambridge University Press, Cambridge (in press).

WOODWORTH, P.L. (1990) A search for accelerations in records of European mean sea level. *International Journal of Climatology* **10**, 129–143.

WOODWORTH, P.L., SHAW, S.M. & BLACKMAN, D.L. (1991) Secular trends in mean tidal range around the British Isles and along the adjacent European coastline. *Geophysical Journal International* **104**, 593–609.

2

Recent sea-level changes

M.J. TOOLEY

Introduction

The Quaternary period (last 2.47 Ma) has been characterised by climatic changes and sea-level changes which have had a profound effect on coastal environments including saltmarshes. In middle and high latitudes, interglacial and interstadial conditions alternated with glacial conditions with the latter more characteristic than the former. In low latitudes, humid and arid conditions alternated. Associated with the glacial/interglacial conditions of middle and higher latitudes, especially in the Northern Hemisphere, were large amplitude and long period changes in sea level. Elevated beaches in formerly glaciated areas provide evidence of a changing relationship between land levels and sea levels. The massive transfer of load from land areas and some continental shelves to the ocean basins of the world during deglaciation had an impact on sea level through adjustments in geoid relief, and both isostatic adjustments and phase changes in the aesthenosphere have also affected this relief pattern over a longer time scale. Over shorter timescales (10 to 1000 years), there is great regional variability of sea-level behaviour which has an important effect on coastal processes.

Sea level can be recorded directly by satellite altimetry or, where it intersects a landsurface, by tide gauges. It can also be recorded indirectly but less sensitively from the geological record. Employing an objective methodology, a data base of sea-level index points can be built up and used to quantify rates of sea-level change and recent vertical earth movements.

Many sea-level problems, both fundamental and strategic, have been addressed in the last 20 years with foci provided by the INQUA Commission on Quaternary Shorelines, the International Geological Programme and the International Geographical Union (Tooley, 1987b). Conference proceedings and edited volumes bear witness to this fertile area of research (e.g. Kidson and Tooley, 1977; Suguio *et al* 1979; Greensmith and Tooley, 1982; Smith and Dawson, 1983;

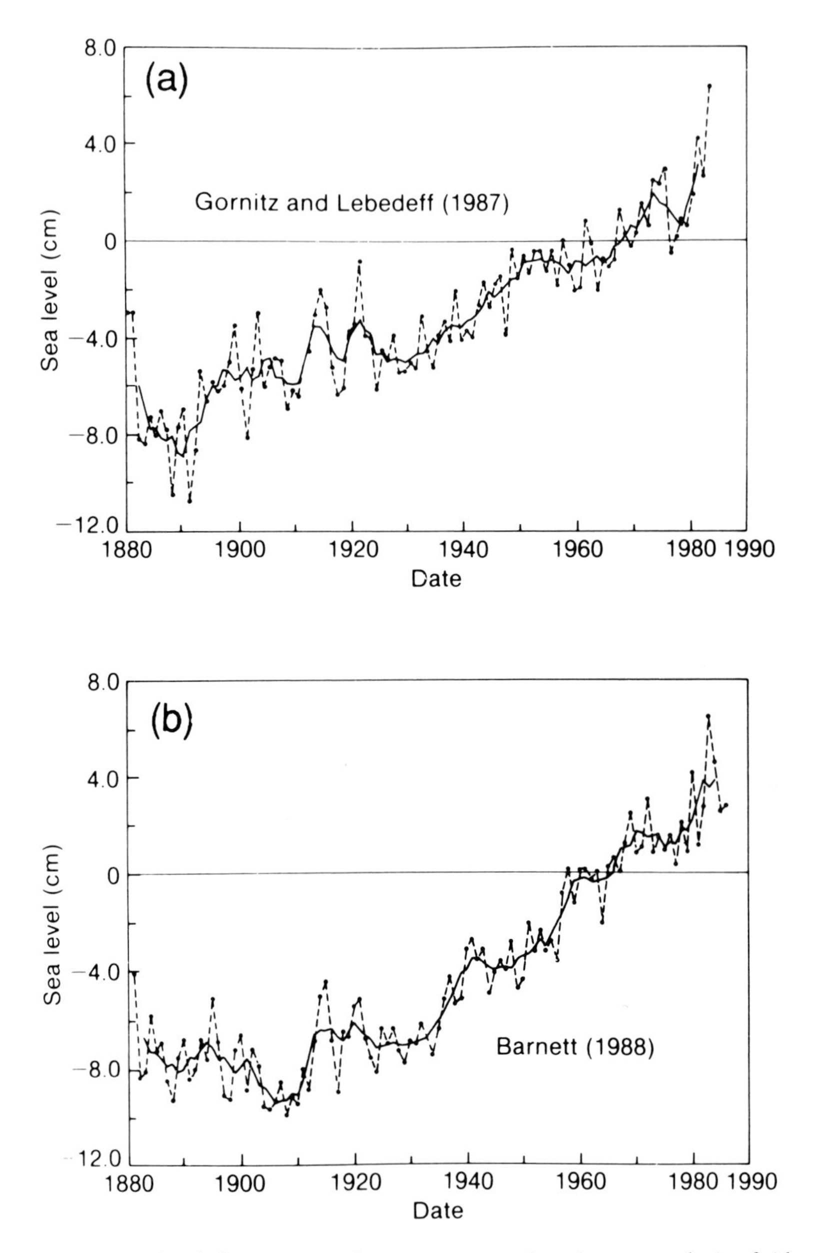

Fig. 2.1. Sea-level changes over the past 100 years based on an analysis of tide gauge data, according to (a) Gornitz and Lebedeff, and (b) Barnett. The average value for the period AD 1951–70 is the zero baseline, and values are plotted in relation to this baseline (from Warrick and Oerlemans, 1990).

Masters and Fleming, 1983; van de Plassche, 1986; Devoy, 1987; Tooley and Shennan, 1987; Shennan, 1989; Tooley and Jelgersma, 1991).

The instrumental record of sea-level changes

Estimates of global sea-level rise during the past 100 years range from 0.5 to 3.0 mm a^{-1}, with most estimates within the range 1.0 to 2.0 mm a^{-1} (Warrick and Oerlemans, 1990). Figure 2.1 shows two solutions arising from analysis of tide gauge data. Gornitz and Lebedeff (1987) analysed data from 130 stations with a minimum record of 20 years, covering the period from 1880 to 1982 with altitudes corrected for recent earth movements, and derived a mean rate of 1.2 $\pm$0.3 mm a^{-1}. Barnett (1988) cited in Warrick and Oerlemans (1990) used data from 155 stations for the period 1880–1986 and produced a mean value of 1.15 mm a^{-1}. These rates of rise have been questioned (Pirazzoli, 1989; Pirazzoli *et al.*, 1989) and it has been argued that if there is a global sea-level (eustatic) signal, it is weak and is obscured by tectonic, glacio-isostatic, anthropogenic and atmospheric effects. Indeed, for the Atlantic coasts of Europe, Pirazzoli (1989) has calculated a sea-level rise rate for the past 100 years of only 0.4–0.6 mm a^{-1} by removing land movements.

By analysing the tide gauge stations with the longest record – Amsterdam, Stockholm, Brest and Sheerness – Woodworth (1990) demonstrated a positive acceleration of mean sea level of 0.44 mm a^{-1} per century, over the past 300 years, and this confirms Ekman's (1988) conclusion from analysis of the Stockholm tide gauge data.

The variability of mean sea level as well as both high and low water has been demonstrated by Van Malde (1991) for the Dutch coast. Whilst a general rise of mean sea level of 1.5–2.0 mm a^{-1} over the last century can be demonstrated, secular variability exists: for example the fall in mean sea level between 1977 and 1987. In addition, harbour works and the closure of the Zuider Zee profoundly affected tidal parameters.

The earliest tidal measurements were made with tidal poles, but the first self-recording tide gauge began operation at Sheerness in 1831, with a float in a stilling well (Pugh, 1990). The location of many gauges is unsatisfactory (for example on quays that have subsided due to sediment consolidation), and the records are often discontinuous. In addition, the gauges frequently fail during storm surges and extreme water levels, when considerable erosion and sediment mobility occur, and reliance has to be placed on temporary marks or modelled solutions.

In 1977, in the Irish Sea, a storm surge added up to 2.5 m to predicted levels, and at Pilling Marsh the extreme water level of +6.86 m OD (Ordnance Datum) intersected and breached the old embankment, causing waterlogging of 2875 ha of agricultural land up to +5.5 m OD and resulting in £750,000 of damage (North-West Water, 1987). At Preston, the same storm surge elevated water levels to +7.5 m OD (Graff, 1978). At the end of February 1990, the demolition of the sea wall at Towyn by waves left a threshold at an elevation of approximately +4.5 m OD, across which subsequent high tides flooded an area of 10 km^2 of residential and agricultural land to a depth of up to 1.8 m. A flood mark in Towyn had an altitude of +5.39 m OD and 2 km south the flood mark was at +5.24 m OD (Englefield *et al.*, 1990; Roe, 1991). The nearest tide gauge at Hilbre Island failed during the storm, but at Alfred Dock, Birkenhead, the tide gauge recorded +6.16 m OD and at Eastham Lock +6.39 m OD (Flather and Proctor, 1990).

Very little sediment was carried landward during the Towyn flood, largely due to the removal of shingle and sand in front of the breached sea wall by longshore drift and the elimination of a source further west by the construction of sea walls. This confirms the conclusion that during storm surges, whilst breaching of natural coastal defences may occur, the landward transport of sediment is restricted (Tooley, 1979). There may be a stratigraphic signature or optical luminescent signal of storm surges in the thick marine clastic layers in the world's coastal lowlands. However, these layers appear to be the consequence of periodic relative sea-level changes throughout geological time rather than aperiodic storm surge events (Tooley, 1979, 1985a).

The geological record of sea-level changes

The stratigraphy of the world's coastal lowlands is similar: there are alternating strata of marine and terrestrial sediments. Sea-level or local factors such as uplift or subsidence, sediment supply, tidal range, catchment size and river discharge will affect the thickness and persistence of these strata. Although there is some variability within and between sites in a single lowland, the sedimentary record is an archive of coastal, sea-level, water level and water quality changes.

Two sites, one in northwest England and the other in Rio de Janeiro State, Brazil, will be described to demonstrate the relationships between lithostratigraphy and sea-level related environmental changes.

Northwest England The coastal lowlands of southwest Lancashire have been the subject of interest and investigation since the seventeenth century. Together with sites further north in Morecambe Bay and southwest in North Wales, they provide a record of sea-level changes during the past 9500 years (Tooley, 1978a). Isostatic uplift in southwest Lancashire appears to have ceased about 6500 years ago (Shennan, 1987), since which time sea-level changes will have been recorded directly in the sediments of the coastal lowlands. The most complete sequence of interdigitating marine and terrestial sediments has been recorded in valleys tributary to the Ribble estuary in the township of Lytham, and shorter sequences have been reported from the intertidal zone, dunes and mosslands around Southport and Formby (Tooley, 1974, 1985a, 1985b, 1987a, 1987b). In this latter area, Downholland Moss has proved to be a key area for sea-level investigations, and one site has been designated a Site of Special Scientific Interest by the Nature Conservancy Council and described in the Geological Conservation Review.

Downholland Moss was adequately drained only in the 1950s, and further improvements occured in the late 1970s. Until very recently, unlike the Fenlands, the organic sediments at the surface have not wasted. Furthermore, sediment consolidation following the recent improved drainage has revealed former tidal creeks with small levees, choked with more recent clastic sediments, comparable to the 'roddons' of the Fenland (Godwin, 1938). The Downholland Moss roddons are associated with the most recent tidal flats which date from 5985 to 5615 radiocarbon years ago. Prior to this, three periods of marine inundation and deposition under tidal flat conditions are recorded (Tooley, 1978b, 1985a), spanning the time period from 8500 until 6200 radiocarbon years ago. Indications of changing environmental conditions are provided by the record of pollen, diatoms and the particle size distribution from one site on Downholland Moss (Fig. 2.2). Here three marine and three terrestrial episodes have been recognized between the surface and a depth of 270 cm.

Marine Episode 1. 271–241 cm. Strata 1–3. The succession from a silt with sand laminations and diffuse iron staining to a clayey silt

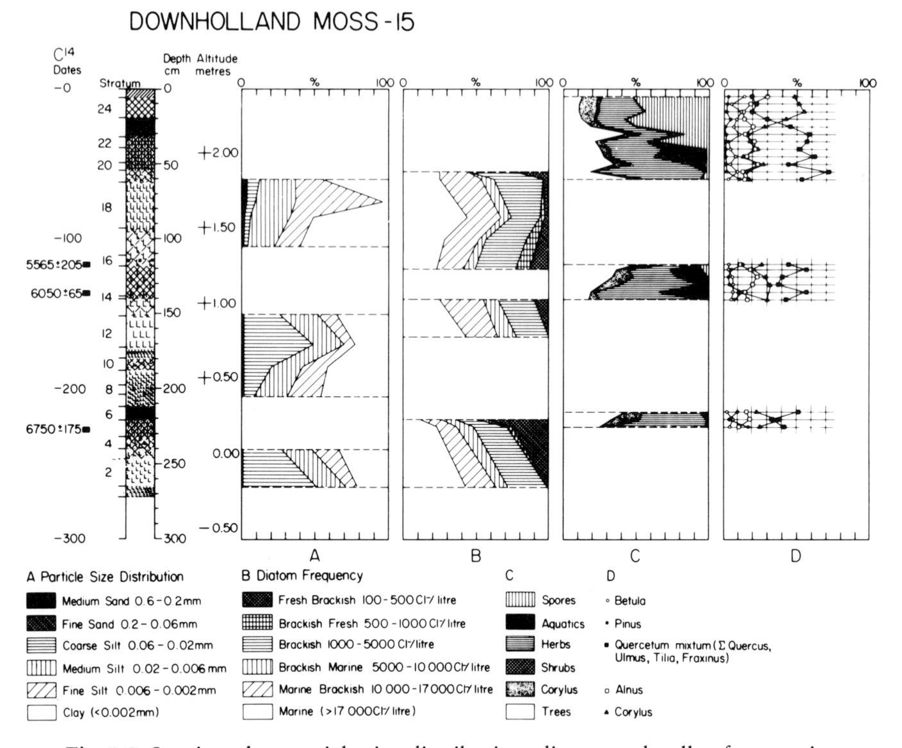

Fig. 2.2. Stratigraphy, particle size distribution, diatom and pollen frequencies in a core from Downholland Moss-15, Lancashire (SD 3202 0838 53° 34′ 02″N. 03° 01′ 37″W). The stratigraphic signatures are according to Troels-Smith (1955). Altitudes are shown in relation to Ordnance Datum (Newlyn) (after Tooley 1978b).

with silt and clay laminations and then to a silt with *limus*, sand laminations and iron-stained root channels points to a change from an upper tidal flat to a saltmarsh palaeoenvironment. The clay fraction increases through this layer, the percentage of marine diatoms declines and that of fresh-brackish water diatoms increases.

Terrestrial Episode 1. 241–213. Strata 4–6. The organic material is *limus detrituosus* with some clay and silt and rare rhizomes of *Phragmites*. The pollen assemblage is characterised by herbaceous taxa. The aquatic taxa in the summary diagram (c) show a decline followed by a rise. *Pediastrum* communities are present indicating freshwater conditions.

Marine Episode 2. 213–140 cm. Strata 7–13. Stratum 7 is a silt with iron staining and discrete *limus* partings, and is interpreted as a saltmarsh palaeosol. This gives way to a silt and silty clay: the presence of stress structures and the absence of laminae points to a palaeoenvironment transition from a saltmarsh to higher mudflats. The frequency of marine diatoms rises and then declines through the strata and through strata 12 and 13 there is an increase in the frequency of fresh-brackish water diatoms.

Terrestrial Episode 2. 140–118 cm. Strata 14–15. A clay *limus* with iron staining in root channels passing up to a slightly-laminated *limus* with occasional sand and silt partings in laminae and *Phragmites* rhizomes. The pattern of the aquatic pollen taxa, dominated by the reed, *Typha angustifolia* is distinctive. From no representation at the 13/14 boundary, they rise to a peak value at 135 cm, before declining. Open fresh water conditions are replaced by oak-alder fen either in response to a falling water table or sediment infill or both processes. Thereafter, the lake reforms, permitting a rise in aquatics, but this peak is not sustained because the lake water becomes progressively brackish and the pollen of saltmarsh taxa such as *Plantago maritima* and Chenopodiaceae begin to rise. This episode lasted for no more than 500 radiocarbon years.

Marine Episode 3. 118–62 cm. Strata 16–18. A silty *limus* with diffuse iron staining and some discrete concentrations passes up into a silty clay with discrete iron partings in sand lenses, horizontally laminated, and iron staining in root channels. A saltmarsh regime is indicated. However, at 75 cm a peak occurs in the frequency of the pre-silt fraction which immediately follows a peak of marine diatoms, and may indicate a shift to higher mudflats. As the 18/19 boundary is approached there is an increase in fresh-brackish water diatoms, heralding a return to organic sedimentation.

Terrestial Episode 3. 62–0 cm. Strata 19–24. A *limus* with Phragmites rhizomes and a *Nymphaea* seed passing up into a crumbly *limus* with sand. In stratum 19, saltmarsh communities give way to fresh, open water communities dominated by the saw sedge, *Cladium mariscus*, which is extinguished by simliar processes to those which occurred in Terrestrial Episode 2. However, in this case, a drying regime is more likely because of the rise in frequency of fern spores and the approach of podsolic dune conditions indicated by the rise in frequency of the pollen of Ericaceae, especially *Calluna vulgaris*. The arrival of blown sand at the western edge of Downholland Moss has been dated to 4090+ 170 (Hv. 4705), and, although almost 1 km west of Downholland Moss-15, described here, some blown sand reached this site; indeed, further east, the limit is about 1.5 km landward.

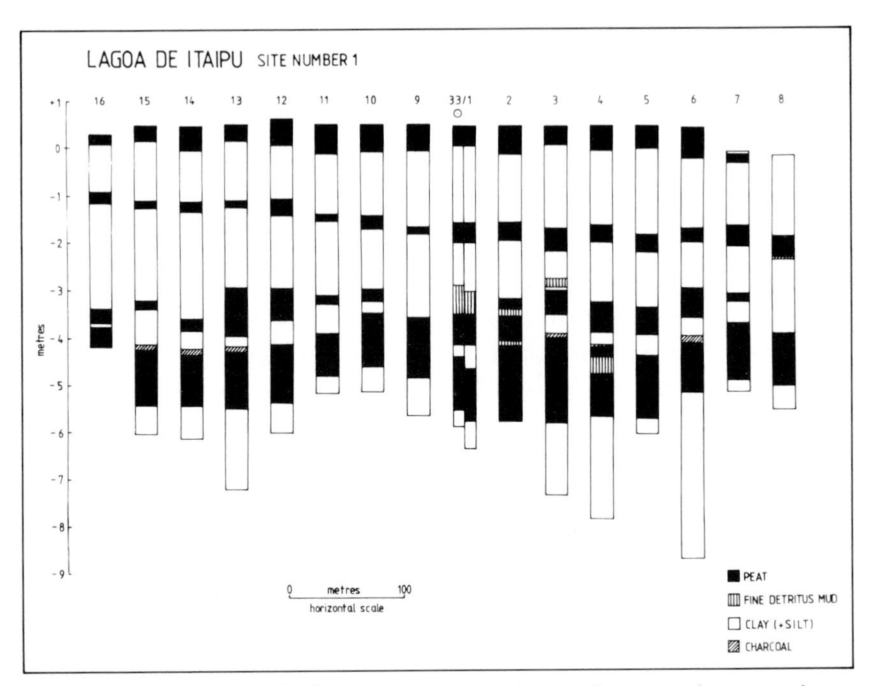

Fig. 2.3. The stratigraphy from a transect running northeast-southwest on the eastern margin of the Lagoa de Itaipu, Rio de Janeiro State, Brazil. Altitudes are shown in relation to the zero levelling datum of Brazil at Imbituba (after Ireland 1987).

Rio de Janeiro State, Brazil The coast and coastal lowlands of Rio de Janeiro State, described by Ireland (1987), contrast markedly with those of Europe in general and northwest England in particular. The faulted escarpment of the Brazilian Shield with monadnock outliers of crystalline rocks is fronted by barrier beaches (*restingas*) of coarse sand and intervening lagoons. There are 20 lagoons along the coast of Rio de Janeiro State ranging in size from 1 to 200 km^2 and in depth from 0.5 to 3.0 m. There are two different height populations of the *restingas* which vary from 4 to 12 m (Muehe, 1979). The lagoons have small catchments which range from 270 km^2 (Marica lagoonal system) to 20 km^2 (Lagoa de Rodrigo Freitas).

An opportunity was provided by a site investigation in the Lagoa de Itaipu, undertaken prior to development, to confirm and elaborate the lithostratigraphy and the stages of infilling the lagoon, in order to obtain sea-level index points from this site on the coast (Ireland, 1987).

Sixteen borings were put down across the eastern shore of the

Lagoa de Itaipu (Fig. 2.3). A basal clay and silt with rare sand of fresh to brackish water origin was overlain by a woody detrital and well-humified herbaceous *turfa* of variable thickness and altitude, in most cases overlain by charcoal and sealed by a fresh to brackish-water clay. This passes up into a second peat, also of variable thickness and altitude. Remarkably, the radiocarbon assays (Fig. 2.4) yielded the equivalent of interstadial or interglacial age dates, and the basal organic and inorganic sediments in this lagoon had survived the transient environmental conditions of low latitudes associated with the mid to high-latitude glaciations. What is also remarkable, if these sequences are interstadial in age, is the marine and brackish water conditions that characterise the basal sequence at altitudes of −6 to −4 m (Imbituba Datum).

Holocene sedimentation begins within the clay and silt stratum above −3 m (Fig. 2.4) which appears to be of freshwater origin, with a brackish water input immediately before the organic layer dated from 7810 to 7110 radiocarbon years ago. This stratum lacks the variability of the lower organic strata and maintains a persistent altitude. Sedimentation was ended by a marine episode which was so sudden at 7100 radiocarbon years ago that some mechanism which includes barrier breaching must be inferred. A long period of dominantly marine conditions ensued, before peat growth recommenced at 370 radiocarbon years ago.

Ireland (1987) has demonstrated that many of the Holocene changes in lithostratigraphy in the lagoons are synchronous, and this implies an overriding control such as sea-level movement.

The methodology of sea-level studies

It is clearly desirable to have a generally accepted, explicit methodology of applied sea-level work in the field and in the laboratory, in order to generate a rigorously tested and refined sea-level data base. Subsequent analyses of rates and directions of sea-level movement and of land movements are possible only from a sound and reliable basis.

Three criteria were enunciated (Tooley, 1978b, 1985a) to select variates that could be used on age-altitude graphs.

The first criterion was that the variates used should come from a small homogeneous area, so that the effects of tidal inequalities, earth movements and variations in geoid configuration would be mini-

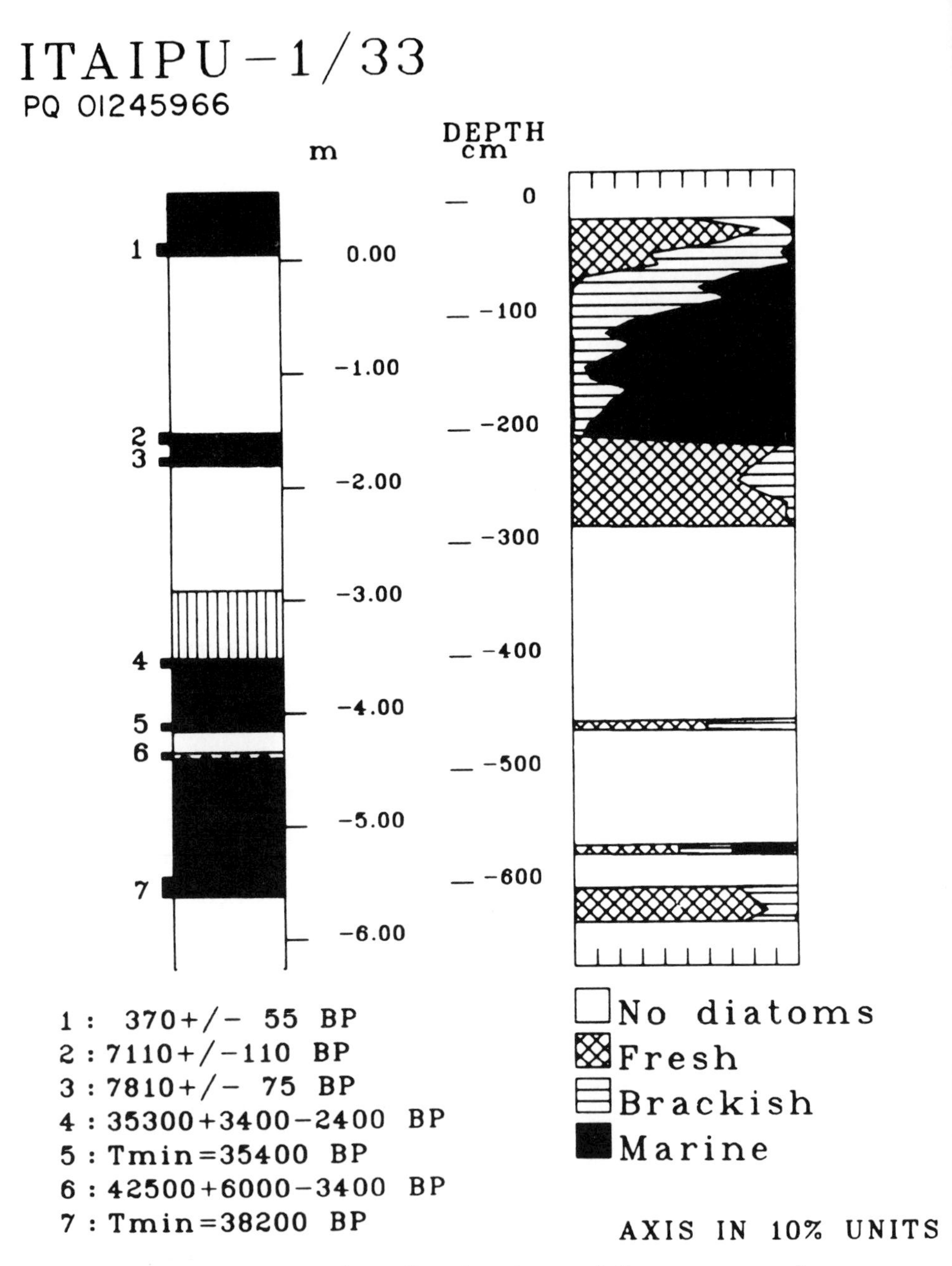

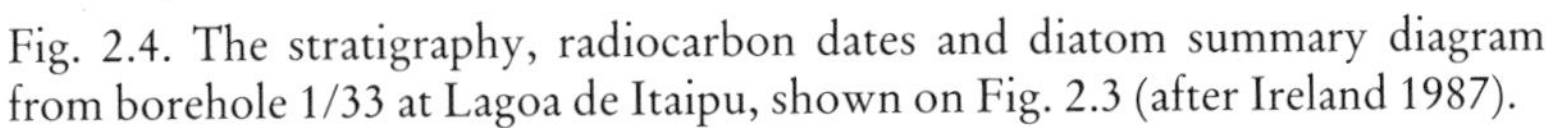

Fig. 2.4. The stratigraphy, radiocarbon dates and diatom summary diagram from borehole 1/33 at Lagoa de Itaipu, shown on Fig. 2.3 (after Ireland 1987).

mised. Subsequent investigations (e.g. Shennan, 1982, 1986a, 1986b; Ireland, 1987; Tooley and Switsur, 1988; Haggart, 1989) have indicated the value of detailed, local scale studies. Within-site variability is recognised and can be accommodated.

The second criterion was that sea-level variates should come from similiar palaeoenvironments and have the same indicative meaning – in other words, the variates should be composed of similar material, the constituent sub-fossil plants and animals of which should be related to a water level, that is ultimately determined by tidal levels on the open coast (van de Plassche, 1986). Each variate should have therefore originated from material that accumulated *in situ* over a very narrow vertical range, which can be related to a palaeo-water level. For example, in temperate coastal lowlands samples of monocoltyledonous *turfa* or *limus* with pollen or seeds of saltmarsh taxa and epiphytic diatoms of marine or brackish water preference (Tooley, 1978a, 1981; Behre, 1986; Palmer and Abbott, 1986) provide ideal material for sea-level index points. Another example would be genera of marine gastropods belonging to the Vernetidae on subtropical or tropical coasts (Laborel, 1986): altitudinal errors range from ±0.1 m on tideless coasts to ±0.5 to 1.0 m where there are strong tides and surf. Coralline algae have been used and are found from the Arctic to the tropics: in the former altitudinal precision ranges from ±5 to 10 m, whereas in the latter algal ridges and trottoirs overlap in close association with sea-level so that altitudinal errors of only ±0.1 m are possible (Adey, 1986). It follows that all sea-level variates should be provided with an altitudinal error band: even tropical coralline algae and marine gastropods should at best be enclosed in an error box.

The third criterion was that radiocarbon dates should be capable of independent corroboration. This can be achieved where standard regional pollen diagrams are available and a chronozone system established. Consistency can be achieved by dating all samples at one laboratory and obtaining several dates from one core. All radiocarbon dates should be displayed with two standard deviations, and together with the vertical errors will provide the dimensions of the error box.

The application of these criteria has led to an overhaul of the nomenclature used in sea-level studies (Shennan, 1982; Tooley, 1982) and the establishment of a sound basis for correlation (Shennan *et al.*, 1983).

Streif (1979) introduced the term 'tendencies of sea-level movement' which was further developed by Shennan (1982). Positive and negative tendencies of sea-level movement can be established from the indicative meaning of the water level indicator. The indicator is site dependent, but indicators from many sites within an area may show a general tendency of sea-level movement, and this is the basis for wider geographical correlations.

Sea-level investigations and the construction of sea-level data bases are considerably aided by the use of a standardised recording system. Whilst others exist (e.g. Streif, 1978), the discipline afforded by the use of Troels-Smith's (1955) scheme has ensured a consistent record. In addition, a preferred route for undertaking sea-level investigations has been proposed (Shennan, 1985).

In the U.K., the application of these criteria under the auspices of UNESCO/IGCP Projects 61 and 200 (Shennan, 1989) has enabled a sea-level data base of 915 records to be assembled. Of these, 487 were peat dates, 78 were wood dates, 81 were shell dates and 85 were dates from prehistoric trackways. Of the 487, considerations of contamination, eroded contact, stratigraphic context and age context reduced this number to 429. For each sea-level index point 67 separate fields of information are required. From this data base, it has been possible to plot sea-level variates on age/altitude graphs by region in relation to Ordnance Datum or a tidal value. In additon, recent earth movements for the United Kingdom or the margins of an epicontinental sea can be calculated (Shennan, 1987, 1989; see Fig. 1.4 in Allen and Pye, Chapter 1, this volume).

Rates of sea-level change

It has been common practice to plot sea-level variates on age-altitude graphs and to fit sea-level curves by eye (for example, Fairbridge, 1961, 1976; Jelgersma, 1961; Jelgersma *et al.*, 1979; Tooley, 1974, 1978a). Whilst this practice has continued (for example, National Research Council, 1987; Fairbanks, 1989), the recognition of all the errors affecting age and altitude, to which a sea-level index point are subject, has led to the use of error boxes, error bands or ellipses for each variate (Kidson, 1982). This practice began in the 1940s (Godwin, 1940), but was abandoned after 1954 (van Straaten, 1954) and not revived until the 1970s (Akeroyd, 1972; Thom and Chappell, 1975; Louwe Kooijmans, 1974). Streif (1989, 1990) has published a banded sea-level curve and concluded that, whilst the band represents the indicative meaning of the sea-level variates, the actual sea-level

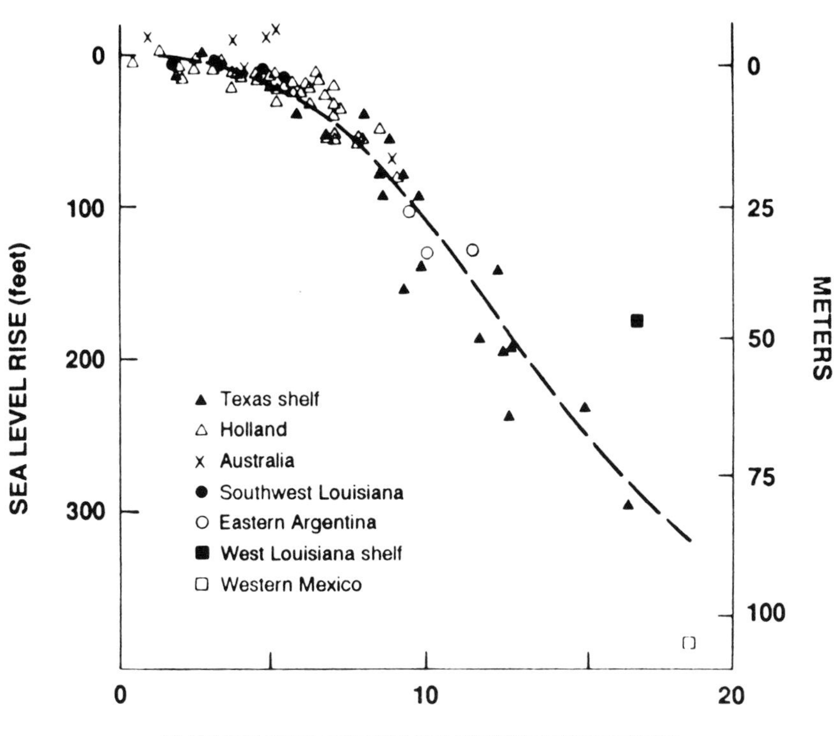

Fig. 2.5. Sea-level variates from 7 locations around the world plotted on an age/altitude graph (after US National Research Council, 1987).

curve lies within this envelope. Trends can be shown but minor oscillations are masked.

Figure 2.5 shows a conventional sea-level curve fitted by eye to variates on an age-altitude graph from seven continental shelf areas and coastal lowlands. Such a practice ignores advances that have been made in sea-level methodology and should be abandoned. Furthermore, Fairbanks (1989) has determined by dating samples of reef crest corals, such as *Acropora palmata*, from Barbados and correcting for uplift, that the maximum lowering of sea level some 18,000 radiocarbon years ago was 121 ±5 m below present levels. Unfortunately, Fairbanks' variates are shown with neither age nor altitude error (cf. Chappell and Polach, 1991). An example of a sea-level envelope on an age-altitude graph is given in Fig. 2.6 from the Fenlands, UK.

Determining rates of sea-level change from banded age-altitude

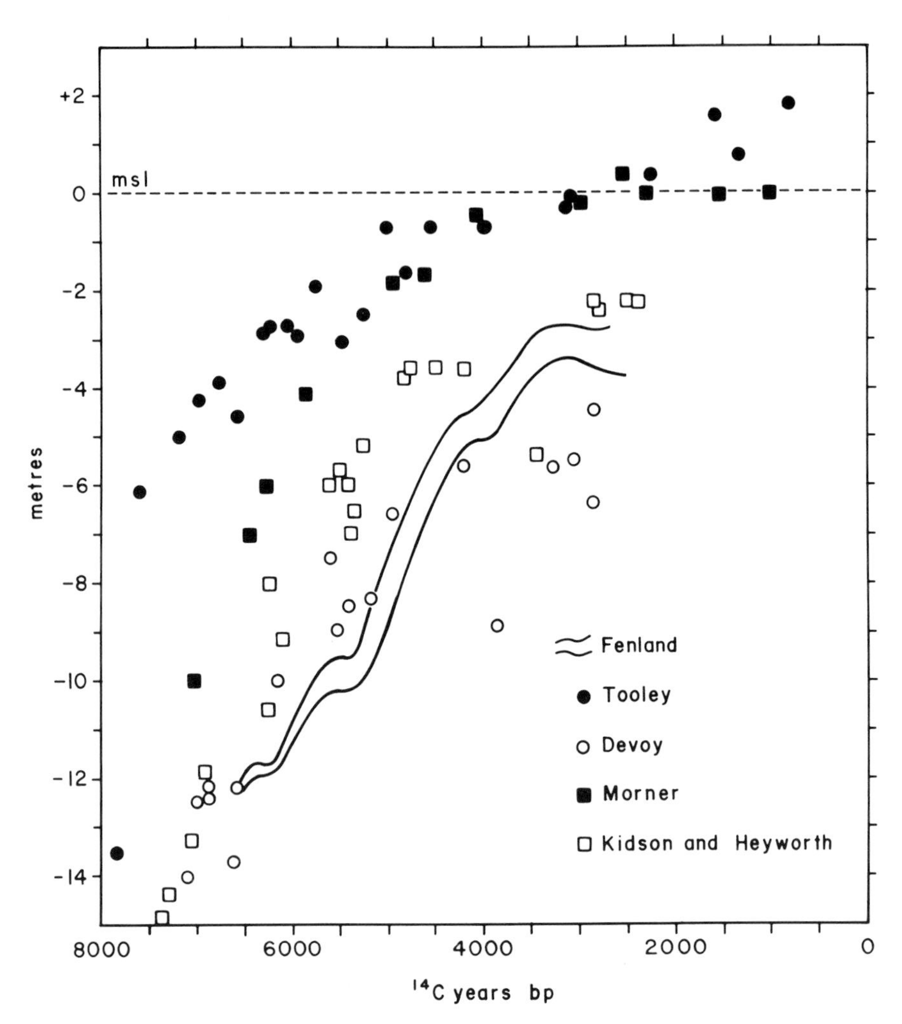

Fig. 2.6. The Fenland sea-level band with altitude and age errors accommodated, compared with sea-level variates from northwest England (Tooley), the Thames estuary (Devoy), the Bristol Channel (Kidson and Heyworth) and the west coast of Sweden (Morner). All values are related to the local Mean Sea Level (from Shennan 1986).

graphs on which sea-level index points are displayed in error boxes should yield for any site a range of values. In practice, sample populations of sea-level variates (radiocarbon-dated samples the measured altitudes of which are known) are considered for short time periods.

There is a generally accepted view that sea-level changes are slow. This view is reinforced by the analysis of the 100 year or so tide gauge records that show a rate of rise of 1.0 to 1.5 mm a^{-1}. This view ignores the evidence from the recent geological past. Cronin (1983) reviewed the evidence for periods of rapid sea-level change and concluded that sea level has risen and fallen at rates ranging from 10 to 30 mm a^{-1}. Tooley (1974, 1978a, 1979, 1987a, 1989) has repeatedly drawn attention to such periods, based on a consideration of the evidence from northwest England. Here, rates of sea-level rise about 7800 years ago range from 34 to 44 mm a^{-1}, and appear to have been driven by the catastrophic melting of the Laurentide ice sheet. Corroboration comes from Denmark (Petersen, 1981) and the German Bight (Ludwig *et al.*, 1981). In eastern China, Yang *et al.* (1987) have indicated periods of enhanced rates of sea-level rise of between 20 and 75 mm a^{-1} with peak rates at about 15,000, 12,500, 11,500 and 7800 radiocarbon years ago. For the late glacial and early postglacial period, Fairbanks (1989) has described two periods of 'exceedingly rapid' sea-level rise, based on coral reef data from Barbados, the first peak at about 12,000 years BP of 24 mm a^{-1} and the second at roughly 9500 years of 28 mm a^{-1}.

Support is given to these enhanced rates of sea-level rise by calculations of meltwater discharge rates, and diversions of meltwater, for example from Lake Agassiz southwards into the Mississippi River and Gulf of Mexico, and subsequently to the St Lawrence River after 11,000 years ago. Discharge rates have been estimated at about 30,000 $m^3 s^{-1}$ and the level of Lake Agassiz dropped rapidly by approximately 40 m (Broecker *et al.*, 1989). Whilst this massive discharge of cold, fresh water disrupted biological productivity, water-mass formation and climate in the North Atlantic, it also had an impact on the rate of sea-level change. Fairbanks (1989) calculated peak discharges of about 14,000 and 9500 $km^3 a^{-1}$ at 12,000 and 9500 years ago respectively. Shaw (1989) concluded that the catastrophic discharge of meltwater from one drumlin field in Canada liberated 84,000 km^3 of water that may have raised sea-level by 23 cm in a few weeks. The simultaneous melting of ice and meltwater discharge from other drumlin fields in the northern hemisphere could be sufficient to raise sea level by several metres in a few years.

Evidence from the last interglacial (Zagwijn, 1983; Streif, 1990) supports the conclusion that sea-level rise was occasionally extremely rapid. During a 450 year period in the Eemian (pollen zones IIIa and early IIIb), the rate of sea-level rise reached a maximum of about 4 mm a^{-1} before slowing to 0.6-0.7 mm a^{-1}.

These latter rates are closer to those calculated for the pre-

Quaternary (Haq *et al.*, 1987). During the Upper Turonian Stage of the Upper Cretaceous the rate of sea-level fall has been calculated to be roughly 0.23 mm a^{-1}.

Conclusion

UNEP's (1990) report on *The State of the Marine Environment* stated that, 'half of the world's population dwells in coastal regions which are already under great demographic pressure, and exposed to pollution, flooding, land subsidence and compaction, and to the effects of upland water diversion. A rise in sea level would have its most severe effects in low-lying coastal regions, beaches and wetlands.'

The forecast for the next century is for a sea-level rise, largely due to thermal expansion of mid- and high-latitude surface water masses in response to global temperature rise, of between 8 and 29 cm by AD 2030 and between 31 and 110 cm by AD 2100 (Warrick and Oerlemans, 1990).

The magnitude and effects of the rise in sea level are likely to be spatially variable. Any addition to the world's oceanic water masses by meltwater will have an impact on the geoid topography. Clarke and Primus (1987) have demonstrated that melting of part of the Antarctic ice cap to yield an equivalent 100 cm 'eustatic' sea-level rise will actually cause a fall in sea level of 200 cm adjacent to part of Antarctica and a rise in sea level of 125 cm in the central Pacific Ocean. In the Thames Estuary, the predicated effect would be a rise of 111 cm and on the Rio de Janeiro coast a rise of 102 cm.

A significant rise in sea level is also likely to affect the coastal wave climate and the frequency and magnitude of storm surges, which will have different regional impacts on the incidence of marine inundation, rates of erosion and deposition, shoreline retreat and advance. Coker *et al.* (1989) have calculated that, if sea level rose by 20 cm, there would be an approximate threefold increase in the number of storm surges equalling or exceeding the danger levels specified by the Storm Tide Warning Service for the east coast of England, based on the period 1972/3 to 1988/9.

The best guide to possible future changes is provided by the sedimentary and palaeoecological record of coastal and sea-level fluctuations during the recent geological past, which provides an insight into local and regional variability. It is important for the planning of land-use changes along the world's coastlines and in coastal lowlands that this variability is recognised and that variable rates of sea-level change are accommodated.

References

ADEY, W.H. (1986) Coralline algae as indicators of sea-level. In O. van de Plassche (ed.) *Sea-level Research*. Geo Books, Norwich, 229-280.

AKEROYD, A.V. (1972) Archaeological and historial evidence for subsidence in southern Britain. *Philosophical Transactions of the Royal Society of London* **A272**, 151–169.

BEHRE, K.-E. (1986) Analysis of botanical macro-remains. In O. van de Plassche (ed.) *Sea-level Research*. Geo Books, Norwich, 413–433.

BROECKER, W.A., KENNETT, J.P., FLOWER, B.P., TELLER, J.T., TRUMBORE, S., BONANI, G. & WOLFI, W. (1989) Routing of meltwater from the Laurentide Ice Sheet during the Younger Dryas cold episode. *Nature* **341**, 318–321.

CHAPPELL, J. & POLACH, H. (1991) Post-glacial sea-level rise from a coral record at Huon Peninsula, Papua New Guinea. *Nature* **349**, 147–149.

CLARKE, J.A. & PRIMUS, J.A. (1987) Sea-level changes resulting from future retreat of ice sheets: an effect of CO_2 warming of the climate. In M.J. Tooley & I. Shennan (eds) *Sea-level Changes*. Blackwell, Oxford, 356–370.

COKER, A.M., THOMPSON, P.M., SMITH, D.I. & PENNING-ROWSELL, E.C. (1989) The impact of climate change on coastal zone management in Britain: a preliminary analysis. Suomen Akatemian Julkaisuja. *The Publications of the Academy of Finland 9/89, Conference on Climate and Water* **2**, 148–160.

CRONIN, T.M. (1983) Rapid sea level and climate change: evidence from continental and island margins. *Quaternary Science Reviews* **1**, 177–214.

DEVOY, R.J.N. (ED.) (1987) *Sea Surface Studies: A Global View*. Croom Helm, London.

EKMAN, M. (1986) A reinvestigation of the world's second longest series of sea level observations: Stockholm 1774–1984. *National Land Survey, Sweden, Professional Papers* 1986 **4**, 10 pp.

ENGLEFIELD, G.J.H., TOOLEY, M.J. & ZONG, Y. (1990) *An Assessment of the Clwyd Coastal Lowlands after the Floods of February 1990*. Environmental Research Centre, University of Durham, 14 pp.

FLATHER, R.A. & PROCTOR, R. (1990) *Notes on the Storms and Coastal Floods in February 1990*. Proudman Oceanographic Laboratory, Internal Document No. 16, 16 pp.

FAIRBRIDGE, R.W. (1961) Eustatic changes in sea level. *Physics and Chemistry of the Earth* **4**, 99–185.

FAIRBRIDGE, R.W. (1976) Shellfish-eating Preceramic indians in coastal Brazil. *Science* **191**, 353–359.

FAIRBANKS, R.G. (1989) A 17,000-year glacio-eustatic sea level record: influence of glacial melting rates on the Younger Dryas event and deep-ocean circulation. *Nature* **342**, 637–642.

GODWIN, H. (1938) The origins of roddons. *Geographical Journal* **91**(3), 241–250.

GODWIN, H. (1940) Studies of the post-glacial history of British vegetation. IV. Post-glacial changes of relative land- and sea-level in the English Fenland. *Philosophical Transaction of the Royal Society of London* **B 230** 285–303.

GORNITZ, V. & LEBEDEFF, S. (1987) Global sea level changes during the past century. In D. Hummedal, O.H. Pilkey & J.D. Howard (eds) *Sea-level Fluctuation and Coastal Evolution.* SEPM Special Publication **41**, 3–16.

GRAFF, J. (1978) Abnormal Sea levels in the north west, *Dock and Harbour Authority* **58**, 366–368, 371.

GREENSMITH, J.T. & TOOLEY, M.J. (EDS) (1982) IGCP Project 61. Sea-level movements during the last deglacial hemicycle (about 15,000 years). Final Report of the U.K. Working Group. *Proceedings of the Geologists Association* **93**, 1–25.

HAGGART, B.A. (1989) Loch Lomond Stadials and Flandrian shorelines in the Inner Moray Firth area, Scotland. *Journal of Quaternary Science* **4**, 37–50.

HAQ, B.U., HARDENBOL, J. & VAIL, P.R. (1987) Chronology of fluctuating sea levels since the Triassic. *Science* **235**, 1156–1167.

IRELAND, S. (1987) The Holocene sedimentary history of the coastal lagoons of Rio de Janeiro State, Brazil. In M.J. Tooley & I. Shennan (eds) *Sea-level Changes.* Blackwell, Oxford, 25–66.

JELGERSMA, S. (1961) Holocene sea level changes in The Netherlands. *Mededeelingen Geologisch Stichtung* C. VI, **7**, 1–100.

JELGERSMA, S., OELE, E. & WIGGERS, A.J. (1979) Depositional history and coastal development in the Netherlands and the adjacent North Sea since the Eemian. In E. Oele, R.T.W. Schuttenhelm & A.J. Wiggers (eds) *The Quaternary History of the North Sea.* Uppsala University Press, Uppsala.

KIDSON, C. (1982) Sea-level changes in the Holocene. *Quaternary Science Reviews* **1**, 121–151.

KIDSON, C. & TOOLEY, M.J. (EDS) (1977) *The Quaternary History of the Irish Sea.* Seel House Press, Liverpool.

LABOREL, J. (1986) Vermitid gastropods as sea-level indicators. In O. van de Plassche (ed.) *Sea-level Research.* Geo Books, Norwich, 281–310.

LOUWE KOOIJMANS, L.P. (1974) *The Rhine/Meuse Delta: Four Studies on its Prehistoric Occupation and Holocene Geology*. Brill, Leiden.

LUDWIG, G., MULLER, H. & STREIF, H. (1981) New dates on Holocene sea-level changes in the German Bight. In S.D. Nio, R.T.E. Schuttenhelm & Tj.C.E. van Weering (eds) *Holocene Marine Sedimentation in the North Sea Basin*. International Association of Sedimentologists Special Publication **5**. Blackwell, Oxford, 211–219.

MASTERS, P.M. & FLEMMING, N.C. (EDS) (1983) *Quaternary Coastlines and Marine Archaeology: Towards the Prehistory of Land Bridges and Continental Shelves*. Academic Press, London.

MUEHE, D. (1979) Sedimentology and topography of a high energy coastal environment between Rio de Janeiro and Cabo Frio – Brazil. *Anais Academia Brasileira Ciencias* **51**, 473–481.

NATIONAL RESEARCH COUNCIL (US) (1987) *Responding to Changes in Sea-level – Engineering Implications*. National Academy Press, Washington DC.

NORTH WEST WATER AUTHORITY, RIVERS DIVISION (1987) *A Guide to the Pilling and Cockerham Tidal Embankment Scheme*. Northwest Water, Warrington, 9 pp.

PALMER, A.J.M. & ABBOTT, W.H. (1986) Diatoms as indicators of sea-level change. In O. van de Plassche (ed.) *Sea-level Research*. Geo Books, Norwich, 457–487.

PETERSEN, K.S. (1981) The Holocene marine transgression and its molluscan fauna in the Skagerrak-Limfjord region, Denmark. In S.D. Nio, R.T.E. Schuttenhelm & Tj.C.E. van Weering (eds) *Holocene Marine Sedimentation in the North Sea Basin*. International Association of Sedimentologists Special Publication **5**. Blackwell, Oxford, 497–503.

PIRAZZOLI, P.A. (1989) Present and near future global sea level changes. *Palaeogeography, Palaeoclimatology and Palaeoecology* **75**, 241–258.

PIRAZZOLI, P.A., GRANT, D.R. & WOODWORTH, P. (1989) Trends of relative sea-level change: past, present and future. *Quaternary International* **2**, 63–71.

PUGH, D. (1990) Sea-level: change and challenge. *Nature and Resources* **26**(4), 36–46.

ROE, G.M. (1991) *Report on the Breach to the Sea Wall in Towyn, North Wales on 26 February 1990*. Seminar on Coastal Protection, 13 March 1991. London, Planning and Transport Research and Computation (International) Co. Ltd. (PTRC), Education and Research Services.

SHAW, J. (1989) Drumlins, subglacial meltwater floods, and ocean responses. *Geology* **17**, 853–856.

SHENNAN, I. (1982) Interpretation of Flandrian sea-level data from the Fenland, England. *Proceedings of the Geologists Association* **93**, 53–63.

SHENNAN, I. (1983) Flandrian and Late Devensian sea-level changes and crustal movements in England and Wales. In D.E. Smith & A.G. Dawson (eds) *Shorelines and Isostasy*. Academic Press, London, 255–283.

SHENNAN, I. (1986a) Flandrian sea-level changes in the Fenland I, the geographical setting and evidence of relative sea-level changes. *Journal of Quaternary Science* **1**, 119–154.

SHENNAN, I. (1986b) Flandrian sea-level changes in the Fenland II, tendencies of sea-level movement, altitudinal changes and local and regional factors. *Journal of Quaternary Science* **1**, 155–179.

SHENNAN, I. (1987) Holocene sea-level changes in the North Sea Region. In M.J. Tooley & I. Shennan (eds) *Sea-level Changes*. Blackwell, Oxford, 109–151.

SHENNAN, I. (ED.) (1989) Late Quaternary sea-level changes and crustal movements in the British Isles. *Journal of Quaternary Science* **4**, 1–94.

SHENNAN, I., TOOLEY, M.J., DAVIS, M.J. & HAGGART, B.A. (1983) Analysis and interpretation of Holocene sea-level data. *Nature* **302**, 404–406.

SMITH, D.E. & DAWSON, A.G. (EDS) (1983) *Shorelines and Isostasy*. Academic Press, London.

STREIF, H. (1978) A new method for the representation of sedimentary sequences in coastal regions. *Proceedings 16th Coastal Engineering Conference, Hamburg*, 1245–1256.

STREIF, H. (1979) Cyclic formation of coastal deposits and their indications of vertical sea-level changes. *Oceanis* **5**, 303–6.

STREIF, H. (1989) Barrier islands, tidal flats, and coastal marshes resulting from a relative rise of sea level in East Frisia on the German North Sea coast. In W.J.M. van der Linden *et al.* (eds) *Coastal Lowlands: Geology and Geotechnology*. Kluwer, Dordrecht, 213–223.

STREIF, H. (1990) Quaternary sea-level changes in the North Sea, an analysis of amplitudes and velocities. In P. Brosche, & J. Sundermann (eds) *Earth's Rotation from Eons to Days*. Springer-Verlag, Berlin, 201–214.

SUGUIO, K., FAIRCHILD, T.R., MARTIN, L. & FLEXOR, J.M. (EDS) (1979) *Proceedings of the 1978 International Symposium on Coastal Evolution in the Quaternary*. USFP, Sao Paulo.

THOM, B.G. & CHAPPELL, J. (1975) Holocene sea levels relative to Australia. *Search* **6**, 90–93.

TOOLEY, M.J. (1974) Sea-level changes during the last 9000 years in northwest England. *Geographical Journal* **140**, 18–42.

TOOLEY, M.J. (1978a) *Sea-level Changes in Northwest England*. Clarendon Press, Oxford.

TOOLEY, M.J. (1978b) Interpretation of Holocene sea-level changes. *Geologiska Foreningen Stockholm Forhandlinger* **100**, 203–212.

TOOLEY, M.J. (1979) Sea-level changes during the Flandrian Stage and the implications for coastal development. In K. Suguio, T.R. Fairchild, L. Martin, & J.M. Flexor (eds) *Proceedings of the 1978 International Symposium on Coastal Evolution*. USFP, Sao Paulo.

TOOLEY, M.J. (1981) Methods of reconstruction. In I.G. Simmons & M.J. Tooley (eds) *The Environment in British Prehistory*. Gerald Duckworth and Co. Ltd, London, 1–48.

TOOLEY, M.J. (1982) Sea-level changes in northern England. *Proceedings of the Geologists Association* **93**, 43–51.

TOOLEY, M.J. (1985a) Climate, sea-level and coastal changes. In M.J. Tooley & G.M. Sheail (eds) *The Climatic Scene*. George Allen & Unwin, London, 206–234.

TOOLEY, M.J. (1985b) Sea-level changes and coastal morphology in North-west England. In R.H. Johnson (ed.) *The Geomorphology of North-west England*. Manchester University Press, Manchester, 94–121.

TOOLEY, M.J. (1985c) Sea levels. *Progress in Physical Geography* **9**, 113–120.

TOOLEY, M.J. (1987a) Long term changes in eustatic sea level. In *International Workshop on Climatic Change, Sea-level, Severe Tropical Storms and Associated Impacts*, Norwich, September 1–4, 1987. UNEP, EC.

TOOLEY, M.J. (1987b) Sea-level studies, in M.J. Tooley & I. Shennan (eds) *Sea-level Changes*. Blackwell, Oxford, 1–24.

TOOLEY, M.J. (1989) Global sea levels: floodwaters mark sudden rise. *Nature* **342**, 20–21.

TOOLEY, M.J. & JELGERSMA, S. (EDS) (1991) *Impacts of sea-level Rise on European Coastal Lowlands*. Blackwell Oxford (in press).

TOOLEY, M.J. & SHENNAN, I. (EDS) (1987) *Sea-level Changes*. Blackwell, Oxford.

TOOLEY, M.J. & SWITSUR, V.R. (1988) Water level changes and sedimentation during the Flandrian Age in the Romney Marsh area. In J. Eddison & C. Green (eds) *Romney Marsh, Evolution, Occupation, Reclamation*. Oxford University Committee for Archaeology, Oxford, 53–71.

TROELS-SMITH, J. (1955) Characterisation of unconsolidated sediments, *Danmarks Geologiske Undersøgelse IV* **3**, 1–73.

UNEP (1990) *The State of the Marine Environment*. Reports and Studies No. 39. UNEP Regional Seas and Studies No. 115.

VAN DE PLASSCHE, O. (ED.) (1986) *Sea-level Research: A Manual for the Collection and Evaluation of Data*. Geo Books, Norwich.

VAN MALDE, J. (1991) Relative rise of mean sea level in the Netherlands in recent times. In M.J. Tooley & S. Jelgersma (eds) *Impacts of Sea-level Rise on European Coastal Lowlands*. Blackwell, Oxford (in press).

VAN STRAATEN, L.M.J.U. (1954) Radiocarbon datings and changes of sea level at Velzen (Netherlands). *Geologie en Mijnbouw* **16**, 247–253.

WARRICK, R.A. & OERLEMANS, J. (1990) Sea level rise. In J.T. Houghton, G.J. Jenkins, & J.J. Ephraums (eds) *Climate Change: The IPCC Scientific Assessment*. Cambridge University Press, Cambridge, 257–281.

WOODWORTH, P.L. (1990) A search for accelerations in records of European mean sea level. *International Journal of Climatology* **10**, 129–143.

YONG HUAIREN, CHEN XIQING & XIE ZHIREN (1987) Sea-level changes since the last deglaciation and its impact on the East China lowlands. In Qin Yunshan & Zhao Songling (eds) *Late Quaternary Sea-level Changes.* China Ocean Press, Beijing, 199–212.

ZAGWIJN, W.H. (1983) Sea-level changes in The Netherlands during the Eemian. *Geologie en Mijnbouw* **62**, 437–450.

3

Saltmarsh geomorphology

J.S. PETHICK

Introduction

Saltmarshes have traditionally been viewed by scientists as depositional systems. Low-energy intertidal zones are often characterised by the accretion of fine-grained sediments, forming mudflats; saltmarshes are the vegetated upper reaches of these. Deposition of tidal suspended sediments is accelerated within the marsh vegetation and the result is that saltmarsh surfaces form quite distinct morphological features on the intertidal profile. The fact that saltmarshes are such distinctive features, that there is such a sharp dividing line between open mudflat and vegetated marsh, and that the processes of deposition are accelerated by the presence of this vegetation has led to the belief that this section of the coast may be examined, uniquely, as a self-contained geomorphological entity and that its morphology may be explained by reference to depositional processes alone.

This paper, however, intends to show that saltmarsh geomorphology is not entirely determined by depositional processes. Deposition, it will be argued, is a necessary, but not sufficient, process to explain the observed morphology of marshes. Yet neither may saltmarsh geomorphology be seen as the result of erosional forces, as, for example, in the terrestrial fluvial system. Instead, the morphology of saltmarshes is the attainment of an equilibrium between stress and strength: the physical expression of the critical erosion threshold. Depositional processes may act up to, but not beyond, such critical thresholds and it is the definition of these that allows explanation of the distinctive saltmarsh morphology.

It is also important to regard the saltmarsh as part of the intertidal profile, rather than as a self-contained landform. There is no reason why low-energy coasts should be seen as possessing any radically different relationship with the forces of wave and tide than high-energy coasts such as beaches. Beaches act as a buffer to waves (Carter, 1988) but so do all coasts, including low-energy mudflats and marshes. Waves on these coasts may be dominated by the semi-diurnal tide wave but the same general relationship holds. This paper, therefore, looks at the geomorphology of saltmarshes as part of the general response of the low-energy intertidal profile to wave energy inputs; a geomorphic response which tends to buffer the energy in such a manner as to resist long-term morphological change. The analogy between mudflats and beaches cannot, of course, be carried too far into the realm of process-response, since the behaviour of cohesionless and cohesive sediments is fundamentally different in many respects, yet the analogy may be helpful at the larger, geomorphological or landform scale, a scale with which this paper is presently concerned.

Saltmarshes and intertidal geomorphology

At the landform scale, the relationship between saltmarsh and fronting mudflat may be viewed as similar, in some respects, to that between sand dune and beach. The sand dune acts as a long-term sediment store: sediment which may be released by high-magnitude, low-frequency events and incorporated in the short-term morphological response of the beach to the imposed energy. Dunes also act as a high-energy wave buffer, dissipating the energy of the highest storm waves which have passed over the beach and therefore allowing reformation of the coastal geomorphology during the subsequent low-energy phase. Beaches which are backed by sand dunes have, consequently, a higher probability of adjusting to, and therefore surviving, rare storm events. A similar analogy may, of course, be drawn between the river flood plain and the river channel; the flood plain acts as an energy buffer during flood events and also as a sediment store which may be released if major morphological response to rare events is demanded.

The geomorphological significance of saltmarshes may be regarded in the same way as these examples, for saltmarshes are a form of insurance policy for the low-energy intertidal profile. They act as an energy buffer in times of storm and are able to release sediments, stored during low-magnitude, high-frequency events, to assist in short-term morphological response of the profile to storms.

Saltmarsh magnitude-frequency response

Very little work has been conducted on the response of low-energy intertidal shorelines to variations in the magnitude and frequency of wave events. Many authors have commented on the fact that saltmarsh deposition can be profoundly influenced by storms, but, in general, this has been to note that coarser grained materials are deposited during high-energy events (e.g. Stumpf, 1983; Reed, 1988).

Measurement of the surface morphology of Essex saltmarshes and mudflats over a five year period has, however, provided an indication of the manner in which these intertidal profiles respond to rare events and therefore may allow some better understanding of their geomorphology.

The Essex coast has mudflat-saltmarsh associations both in exposed open coast locations (as on the Dengie Peninsula) and within the protection of estuaries (e.g. the Rivers Blackwater and Crouch) or bays (e.g. Hamford Water). The tidal range on this coast averages 4.5 m and this is the dominant energy input into the coastal system within the most protected estuaries and bays, but, on the open coast and for some distance into the estuaries, wind-generated wave energy can be high and dominates the coastal geomorphology.

Wave recording on the coast and in the Blackwater estuary during the period 1987-90 showed that a severe storm period was experienced during the winter of 1989 (Fig. 3.1) when onshore easterly winds caused waves whose maximum significant height attained 3.4 m at the saltmarsh edge. During this period intertidal surface elevation was being monitored using a series of buried plates as reference planes. Figure 3.2 shows that the response of the most exposed of the intertidal profiles to the storm event was dramatic: the saltmarsh eroded vertically while the fronting mudflat accreted. (The data are average figures for mudflat and saltmarsh transects). Simultaneously, the vegetated edge of the saltmarsh retreated horizontally by 5 m. During the period before and after the storm events the marsh accreted steadily at a rate of 1.4 cm per year while the mudflat eroded by a similar amount.

These observations demonstrate that the response of this low-energy intertidal profile to wave events is exactly as would have been predicted by classic beach theory. As long ago as 1940 the work of Shepard and Lafond (1940) demonstrated that beach profiles flatten during storm events and steepen during intervening low wave energy periods. Since then the work of the Australian school of coastal geomorphologists has demonstrated that this response is one which allows maximum energy dissipation during high-energy storms and maximum energy reflection during calm periods. The observed beach profile flattening during storms is, of course, merely another way of describing an increase in the length of the intertidal profile, thus

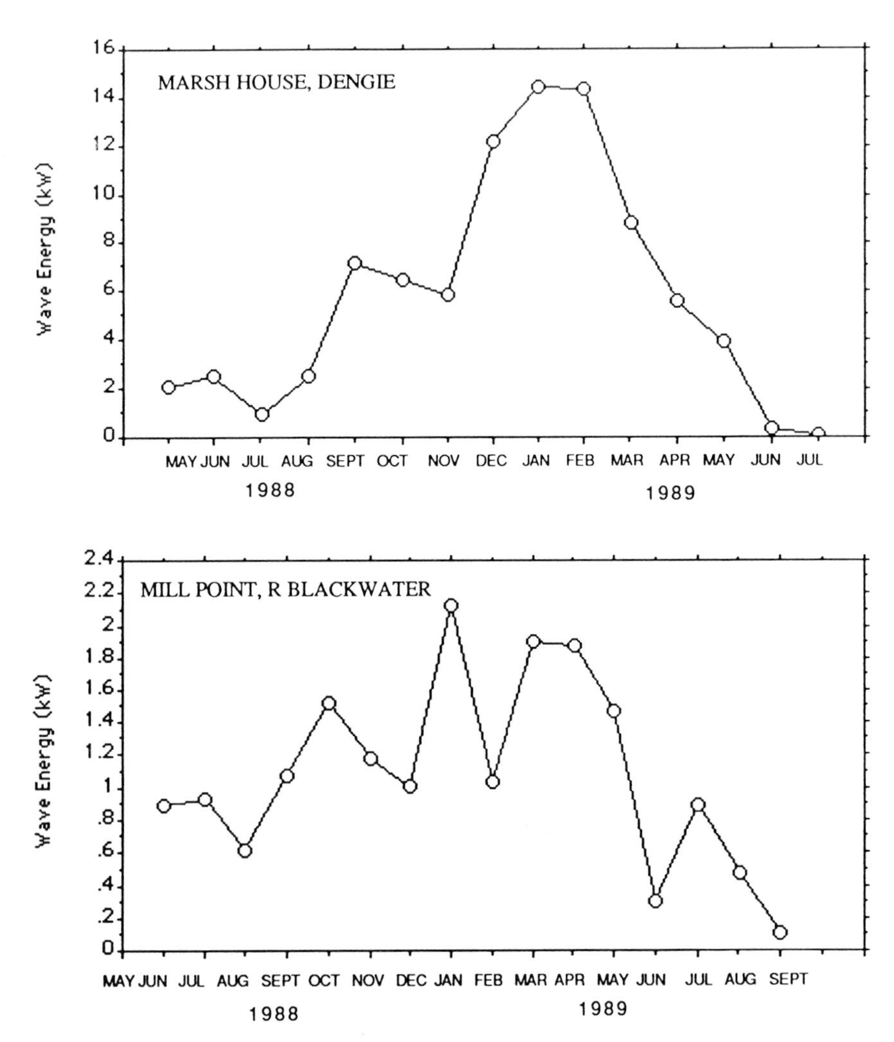

Fig. 3.1. Annual variation in wave energy at (top) an open coast marsh in Essex and (bottom) an estuarine marsh (R. Blackwater, Essex) during 1988–9.

allowing wave energy to be dissipated over a much longer distance and maximising energy dissipation.

The response of the Essex mudflat-marsh system followed this classic coastal response. During the storm event the decrease in average marsh elevation and increase in average mudflat elevation resulted in a flattening of the whole intertidal profile. This also implies an increase in the length of the profile but, since the Essex saltmarshes here are backed by a sea wall, an overall increase in the length of the profile may not have occurred. Instead the mudflat

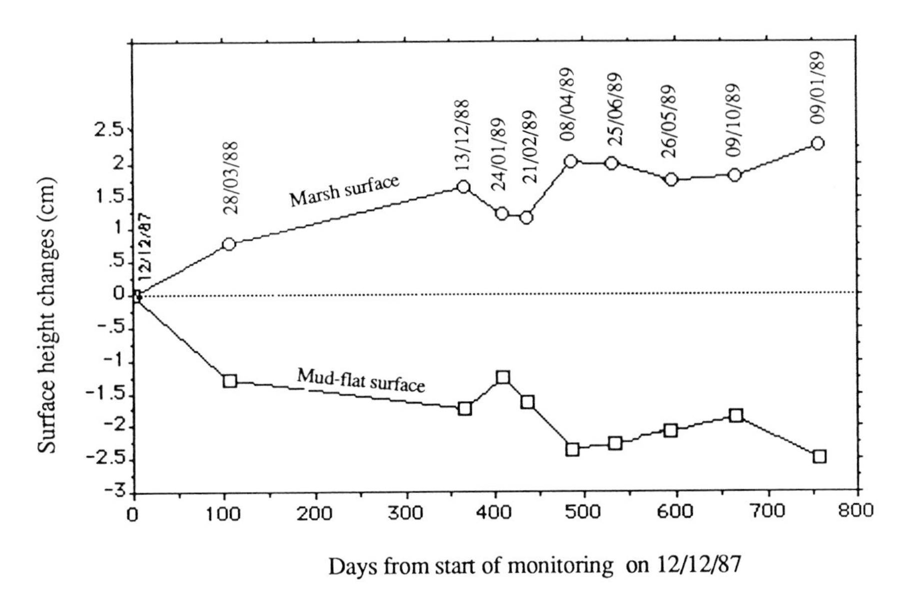

Fig. 3.2. Vertical changes in surface height, open coast marsh, Dengie Peninsula, Essex. Note effect of major easterly storms from 13.12.88 to 21.2.89.

section of the profile lengthened by eroding into the saltmarsh edge. The similarities between the behaviour of the mudflat-saltmarsh profile and that of beach-sand dune to storm events is so marked as to suggest that they represent the same general geomorphological response.

A similar response to the wave energy conditions during the period 1987–90 was shown by a more protected intertidal profile some 10 km within the River Blackwater. Here wave energy was diminished due to refraction and friction within the estuarine channel but, nevertheless, waves of 1.5 m at the marsh edge were experienced during the 1989 storms. The profile again flattened and increased in length as described above (see Fig. 3.3), saltmarsh surface erosion being matched by mudflat accretion and erosion of the vegetated marsh edge.

In both the open coast marshes and those of the protected estuary environment, the almost perfect mirror images exhibited by the changes in elevation of saltmarsh and mudflat shown in Figs 3.2 and 3.3 suggest that sediment is interchanged between the two zones of the intertidal profile. Thus the storm erosion of the saltmarsh surface provides sediment for the accretion of the mudflats. Such a conclusion is supported by measurements of suspended sediment movement

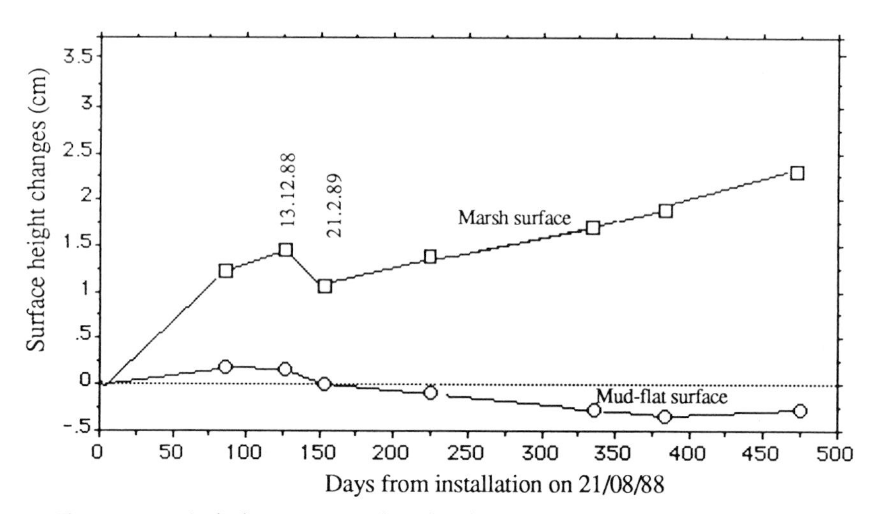

Fig. 3.3. Vertical changes in surface height, estuarine marsh, Mill Point, R. Blackwater. Note effect of storms between 13.12.88. and 21.2.89.

during a tidal cycle (Pethick, 1991). Flood tide currents resuspend sediment laid down during the previous ebb tide on the lower mudflats. This suspended sediment then moves upwards and is deposited on the upper mudflats and saltmarshes during the high water period. As the tide turns, so the deposits on the upper mudflat sediment are resuspended and are carried to low water where they are again deposited during the slack water period. Erosion of either the saltmarsh or mudflats, therefore, results in an increase in the suspended sediment concentration of the 'pool' of sediments kept within this intertidal circulation and is available for deposition, either for short-term residence on the mudflats or long-term residence on the saltmarshes.

The observations described above indicate that the intertidal mudflat-saltmarsh geomorphology is a response to wave energy. Such a conclusion is, of course, hardly an original one when applied to the intertidal profile generally. Anderson (1972) has shown that mudflat morphology can be explained by the stress-strength relationships at the mud-water interface. What is significant about the results of the Essex work, is the fact that the saltmarsh appears to act as an integral part of the profile response, storing and releasing sediments and acting as a morphological extension to the mudflat profile during storms. These observations provide an important pointer towards an explanation of the more detailed morphology of the saltmarsh itself: a pointer which may be followed by first examining the process of morphological recovery of the Essex saltmarshes from the storm events of 1989.

Saltmarsh recovery from storm events

The saltmarsh surface The return interval for the most extreme wave which affected the Essex saltmarshes during 1989 was 1 in 5 years ($p < 0.00006$). This is, however, the return interval of the wave at sea; on the saltmarsh surface the wave can only have effect during high water spring tides. Assuming the marsh is covered by the tide for 4 hours on each of 5 days during the lunar cycle the probability during a year of obtaining a marsh tide is $p < 0.03$. Multiplying these probabilities indicates that the return interval for this wave on the saltmarsh surface is 1 in 33 years.

Observations during the remaining period of the 5 years that the work has progressed shows that no other erosional event took place on the saltmarsh surface and thus the critical return interval for a wave which will erode the marsh surface must lie between 1 in 5 and 1 in 33 years. Although further research is clearly required here before more precise definition is possible, these initial results do indicate that the vertical erosion of the saltmarsh surfaces during 1989 can be regarded as a rare event. The implication is that the surface geomorphology of the saltmarsh is not necessarily formed by such erosional events since the processes of deposition may allow recovery of the surface.

Figure 3.4 shows that this is indeed the case. The average elevation of the saltmarsh was lowered by 0.6 cm during the storm event. From October 1989, however, the surface began to recover from this erosion and deposition was constant thereafter. By early 1990 the surface had returned to its pre-storm elevation while by the autumn of 1990 the surface had risen to 0.4 cm above the pre-storm level. It would appear that the marsh has recovered remarkably quickly within 2 years, but this ignores the fact that deposition would have continued unchecked for those 2 years had erosion not taken place. Figure 3.5 shows that the rate of deposition increased slightly due to the fall in surface elevation after the erosional event. This meant that the surface elevation rose more quickly after erosion had ceased than it would have done pre-erosion, so that the surface will eventually 'catch up' with the extrapolated marsh growth. Figure 3.5 shows that this process of catching-up will take 5 years.

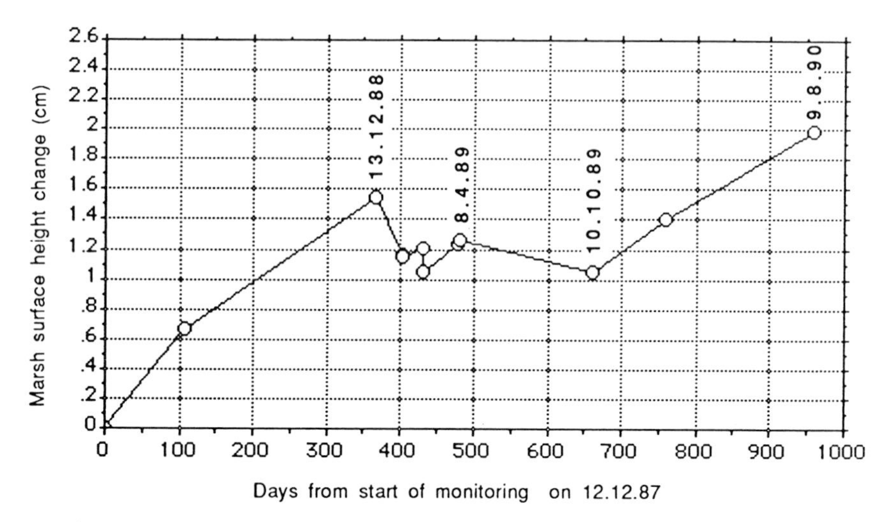

Fig. 3.4. Mean change in saltmarsh surface elevation, 1987–90. Open coast marsh, Dengie Peninsula, Essex.

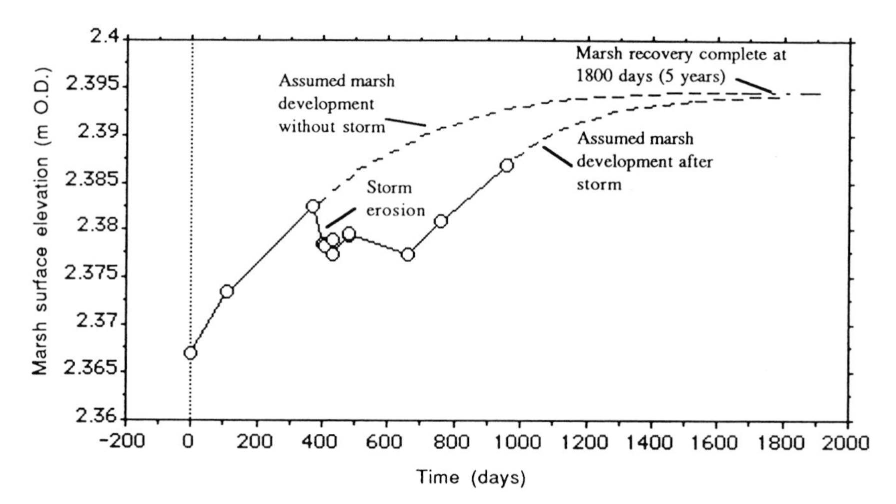

Fig. 3.5. Theoretical recovery of saltmarsh after storm erosion. Based on data set shown in Fig. 3.4.

Such an argument indicates that erosion must affect saltmarshes at intervals greater than 1 in 5 years, since if the rate was any greater than this the marsh surface would not be capable of 'catching-up' and, therefore, saltmarshes would disappear and, indeed, would never have developed in the first place. Is is clear that these results apply to one marsh at one stage in its development. Return intervals will be different elsewhere and will vary as the marsh elevation rises at a given location. More work is required to define such variability more precisely.

The saltmarsh edge If the saltmarsh surface can recover from erosional events by acclerated deposition, then the morphology of the saltmarsh surface must indeed be determined by depositional processes. Pethick (1981) and Allen (1990) have argued that deposition rates will show an exponential rate of decrease as marsh elevation increases so that an upper asymptote may be neared but never attained. Some controversy exists as to the determinants of the elevation of this asymptote (see, for example, French *et al.*, 1990 and Pye, chapter 8, this volume) but these do not preclude a general conclusion that the marsh surface rises towards an upper limit, provided that the organogenic content of the marsh sediments is low and that terrestrial inputs are negligible (Allen, 1990). The marsh profile, therefore, eventually attains a horizontal upper limit and its development through time is one of continuous flattening interrupted by occasional erosional events.

The fronting mudflat, on the other hand, experiences an upper limit to its deposition which is governed by either wave or tidal current shear stress so that the mudflat profile does not progressively rise and flatten but retains a significant seaward slope. The much higher initial accretion rates on the saltmarsh compared with those on the mudflat, together with the contrast in the upper limits to such deposition in each case, mean that the interface between mudflat and marsh is marked by a break in slope which often develops into a vertical cliff 1 m or more in height. That such marsh cliffs can be a response to depositional processes alone is always difficult to accept, since they exhibit all the forms of larger-scale erosional sea-cliffs. Indeed, erosion can in many cases be observed directly and a purely depositional development is made even more difficult to accept.

The observation of the response of the Essex coast to storm events, described above, creates an even more complex problem facing a coherent explanation of marsh cliffs. During the 1989 storms the vegetated edge of the Essex saltmarshes was observed to recede. The explanation given above was that such recession allowed the mudflat profile to increase in length, and thus increased wave energy dissipation. The question then arises as to the manner in which the marsh edge may recover from this erosional event. Since larger magnitude wave events will always occur, then erosion of the marsh edge must continue intermittently and, since it is matter of observation that saltmarshes still exist, the marsh edge must recover from such erosional events. There are several hypotheses which may be advanced to explain how such recovery may take place.

In the case of *protected marshes*, the basic premise stated above that larger magnitude wave events will always occur does not apply universally. Only those intertidal areas experiencing wave energy inputs from the open sea, or bodies of water where fetch lengths are not limiting, will suffer such an open-ended wave probability distribution. Coastal areas which receive wave energy from fetch-limited water bodies will have wave probability densities with an upper, finite, limit imposed by the fetch length available. Many inner estuarine marsh areas come into this category: since they are totally protected from the sea they experience only waves generated within the estuary itself Such waves may be both fetch- and duration-limited and, consequently, have a low upper height limit. Saltmarshes in such protected areas may therefore experience waves which never attain heights greater than a few centimetres. In such cases the marsh cliff may be in equilibrium with this upper wave limit: the marsh advances forwards until the critical stress-strength threshold at the cliff face is attained.

Turning to *open coast marshes without cliffs*, that is, those which experience an open-ended wave probability density, cliff erosion must occur at intervals. The energy dissipation properties of the fronting mudflat may lower the periodicity of such events but cannot prevent erosion completely. If the marsh survives in such conditions it is clear that edge recovery must take place, just as the surface recovers vertically as described above. If the marsh edge is not cliffed then such recovery poses no problem. During periods of sub-critical wave energy, the marsh vegetation advances forward over the fronting mudflat. Supra-critical wave events then erode this edge, perhaps at a frequency of 1 in 5 years or longer according to the Essex data set given above. The marsh then recovers from such erosion by readvancing once more. Marsh-mudflat systems which operate in this manner have no break in slope between them; they are usually younger marsh systems which have not yet attained their final surface elevation but mature marsh systems can be observed facing open coasts which do not possess marsh cliffs as, for example, on the Lindisfarne mainland marshes.

Many open coast marshes are characterised by a highly dissected edge on which vegetation growth is absent or ephemeral. The marshes of Essex and those of the outer Humber Estuary are examples of this development which, in Essex, has been termed 'mud mounds'. The dissected edge consists of a series of rills cut deeply into the intertidal profile surface. These rills develop into intricate branching networks (Fig. 3.6) and are formed by wave action. Their geomorphological significance is probably that they dissipate medium range magnitude-frequency wave events by forcing the waves into the intricate system of hydrodynamically inefficient channels in which their energy is both sub-divided and dissipated in friction, thus remaining below the critical erosion threshold. The rills do not, of course, dissipate high-magnitude wave energy and during these events the marsh edge itself is eroded back as described above. The marsh edge, however, exists at two levels: the rill bottoms and the interfluves between them. The interfluve surface is contiguous with the marsh and the fronting mudflat without marked break in slope and certainly without any cliff development. Thus, during storms, the vegetated edge of the marsh retreats and the surface elevation falls due to vertical erosion, but no cliff development takes place. Recovery of the marsh edge is consequently uninhibited by a cliff and readvance of the vegetation over the upper surface of the mudmounds takes place.

Finally, we consider the case of *open marshes with cliffs*. Where mature marshes face the open sea or wide estuaries, then the presence of a cliff together with an open-ended wave probability density is most difficult to explain. One possible explanation is provided by Allen (1989) in his detailed description of the marshes of the outer Severn estuary. Here the development of a secondary marsh, below the upper marsh and separated from it by a marked cliff, may allow the periodic advance and retreat of the outer marsh which is contiguous with the fronting mudflats without affecting the marsh cliff itself. Such secondary marshes are widely distributed, sometimes forming paired terraces within the wider saltmarsh creeks.

The observed erosion of marsh edges in many areas of the world does not necessarily preclude the discussion given above. Observed erosion may, of course, be a response to a rare storm event and subsequent recovery may occur. Rising sea levels in many marsh areas means that the periodicity of wave events is increasing so that all marsh edges will be forced to retreat in response to increased wave energy. This may apply even to marshes located in wave fetch- and duration-limited areas, since here rising sea levels will increase both the duration of inundation and the areas of water available for wave generation.

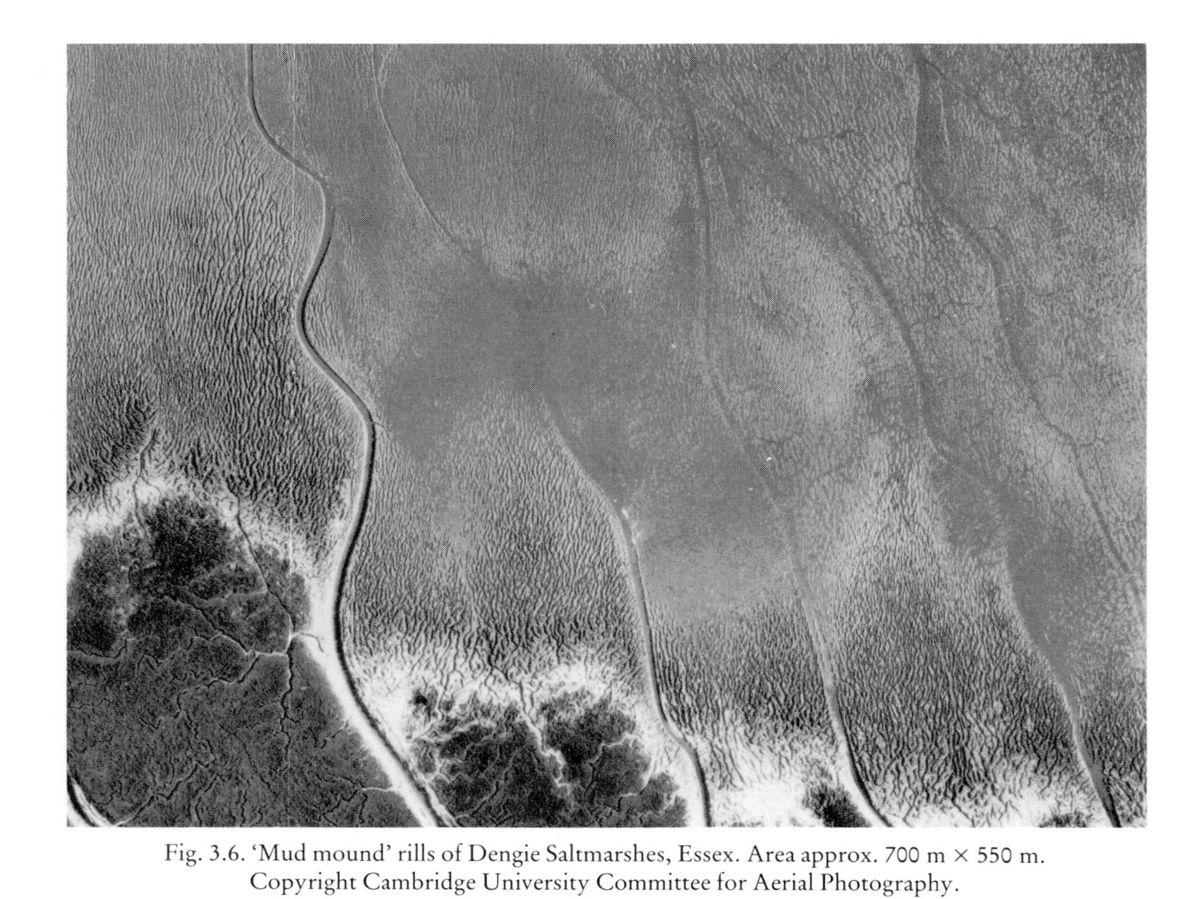

Fig. 3.6. 'Mud mound' rills of Dengie Saltmarshes, Essex. Area approx. 700 m × 550 m.

Another explanation for observed erosion of marsh edges is that extrusion of sediments by compression within the marsh may force sediments forward along the cliff edge. These extruded sediments may then be eroded by storm waves allowing the marsh to retreat and advance in response to energy conditions. This process may also be responsible for the apparent erosion of creek banks within the marsh which may be observed directly, yet which is shown by map and air photograph evidence to result in neither creek widening nor re-orientation.

Saltmarsh creeks

The saltmarsh-mudflat system has been described above as one which acts as a buffer to wave energy and which has developed a series of morphological characteristics in response to periodic wave events. The wave energy inputs discussed so far, however, have been those generated by winds, either on the open sea or within the enclosed waters of estuaries or saltmarsh creeks. Tidal energy inputs into low-energy coastal systems, however, play an important part in the development of mudflat and saltmarsh geomorphology. Where such coastlines face the open sea, the power (i.e. energy per unit time) input during storms from wind-generated waves is far in excess of tidal power inputs; high-frequency tidal inputs, nevertheless, play an important role even on these open coasts. In the more protected environments of estuaries and bays, however, tidal energy inputs are the dominant geomorphological force acting, since here, not only are wind-generated waves limited in height, but also tidal energy is enhanced by a number of resonant factors producing large tidal ranges which flow in estuarine channels producing high current velocities. Mudflat response to such estuarine tides is complex and it is not the purpose of this paper to attempt any detailed explanation of their morphology. However, it appears that the dominant intertidal profile shape of small tidal estuarine channels is convex upwards and that this is a response to the mid-tide velocity maximum followed by high tide slack water, a characteristic of partially reflected tides in channels.

Several authors have commented on the plan shape of the unconstrained, coastal-plain estuary. Thus Langbein (1963), Myrick and Leopold (1963), Leopold *et al.* (1964) and Wright *et al.* (1973) all note that such estuaries exhibit an exponential decrease in width inland and that this morphology acts in such a way as to force the tidal flow into a progressively inefficient channel cross-section so that tidal energy is dissipated through friction. Leopold *et al.* (1964) go further and suggest that, for a branched tidal estuary, the total width of all channels at incremental distances from the estuary mouth will exhibit an exponential decrease with distance inland. Such a branched channel will sub-divide the flow and allow more rapid energy dissipation than an unbranched estuary.

Evidence from the North Norfolk saltmarshes shows that the sum of the width of all creeks at incremental distances from the creek mouth does show exponential decreases inland (Pethick, in press). This means that creek width decreases dramatically at each successive bifurcation and may account for the fact that the most landward of saltmarsh creeks are so narrow as to be covered by vegetation at the surface even though they may be 1 m deep or more. It also means that the total length of saltmarsh creeks is dependent on the width of the mouth of the system. Thus the North Norfolk marshes exhibit creek systems which end abruptly some distance from the landward edge of the marsh leaving the back marsh without any channel system. Most of these North Norfolk marshes have roughly similar mouth dimensions although widely differing marsh surface areas and, as a consequence, the creek density (total creek length/marsh area) is inversely related to marsh area in these marshes. This implies that creeks are not a morphological response to drainage of the ebbing tide from the marsh, in which case creek density would be independent of marsh area, as it is for terrestrial drainage systems (Pethick, 1978). Instead, the morphology and dimensions of the creek system are dependent on the tidal prism which enters the marsh during the flood phase. The development of the marsh creeks may therefore be seen as a morphological device to dissipate tidal wave energy, in precisely the same manner as described for large tidal estuaries by Wright *et al.* (1973).

It is important here to stress that, although the creek system is defined by the saltmarsh edge, the tidal processes which govern their morphology are largely independent of the marsh and may rather be regarded as part of the mudflat system which acts as the principal wave buffer of the system, in the same way as the beach acts in the high-energy coastal system. The creeks are, therefore, more an extension of the mudflat into the saltmarsh than an integral part of the marsh; moreover, once such a distinction is made, many of the apparently obscure aspects of marsh creeks can be explained.

Although a complete examination of the morphological variability of saltmarsh creek systems is not possible here, it may be useful to examine the range of creek morphology present in the Essex and Norfolk marshes.

The Essex open coast marshes The marshes of the open coast in Essex have already been described in this paper. The mud mounds of the saltmarsh edge, for example, were described as a morphological device which forced the wind-generated waves into hydrodynamically inefficient channels, thereby dissipating wave energy. This is exactly the process which, it is now suggested, applies to the larger-scale creek systems.

Figure 3.7 shows some of these creek systems on the Dengie Peninsula saltmarshes. The creek systems appear to have three distinct morphological zones. First, the seaward section of each of the main creeks can be seen to consist of a wide mouth (~50 m) rapidly decreasing in width to no more than ~5 m over a distance of ~100 m. Further inland, the second morphological zone is characterised by a long, parallel-sided creek with few tributary junctions. These extended creeks reach practically to the back of the marsh where they terminate in a series of tributary branches so that the creek system develops its third morphological zone – one characterised by an intricate branched creek network.

Only the first of these morphological zones corresponds to the pattern of exponential decrease in width which characterises larger tidal estuaries. The long parallel-sided middle section is quite unlike other creek systems or estuaries. If the outer section is a morphological response to tidal energy dissipation then the rapid decrease in width suggests that tidal energy here is not powerful enough to extend more than 100 m or so into the marshes. Tidal energy on the open coast is, indeed, limited: the rectilinear tide flow is shore parallel, so that only very low-energy inputs are directed towards the marshes. On the other hand, the middle morphological zone, with long parallel-sided creeks, indicates a much more energetic mechanism at work. It may be that these creeks are a response to low-frequency but high-magnitude wind-wave energy, forcing water into the tapering mouth of the outer zone which fails to dissipate this energy. The wave energy thus forces further into the marshes, creating the elongated creeks. Thus the middle morphological section may be seen as an extension of the mudflat into the marshes on which waves as high as those recorded in 1989 (3.5 m) can be dissipated. Once most of this wave energy has been dissipated in the long middle

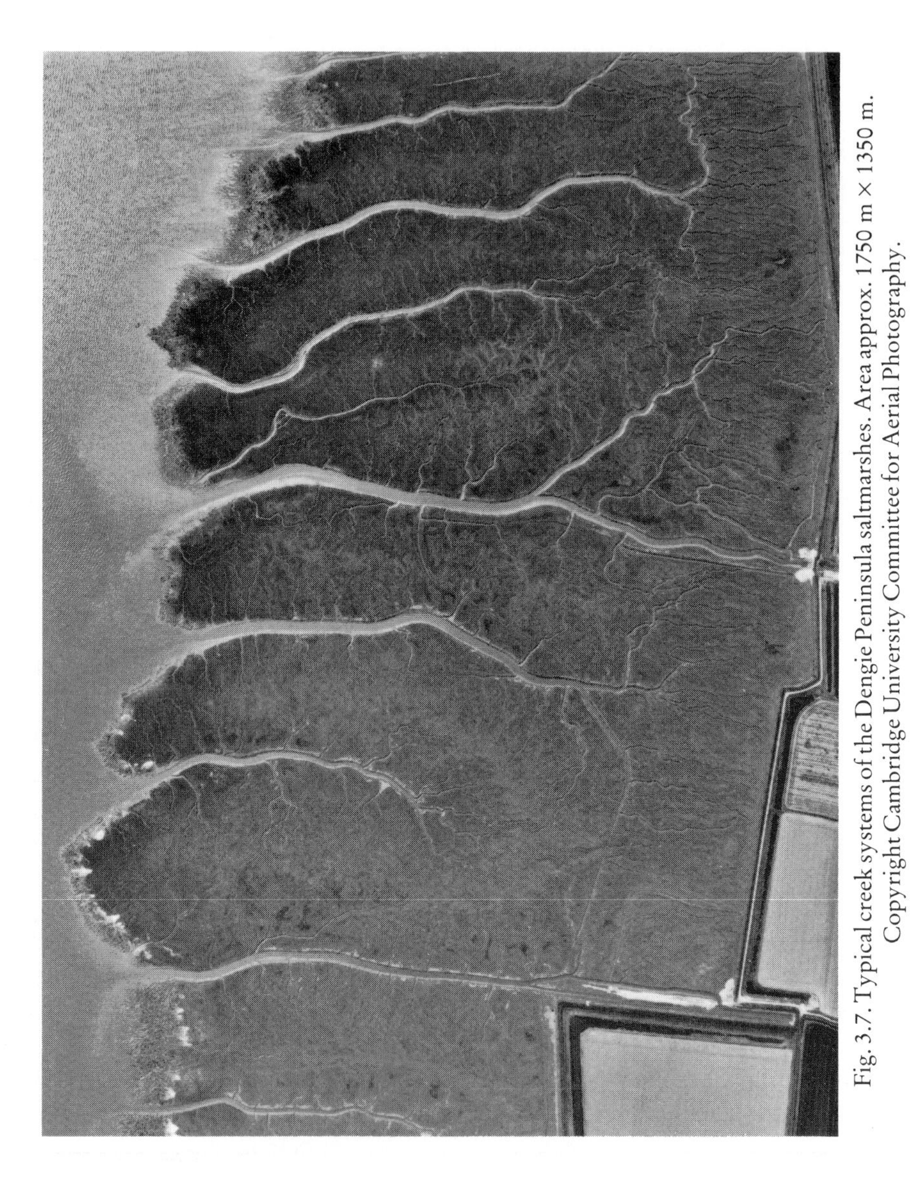

Fig. 3.7. Typical creek systems of the Dengie Peninsula saltmarshes. Area approx. 1750 m × 1350 m. Copyright Cambridge University Committee for Aerial Photography.

section of the channel, the creeks bifurcate, allowing sub-division of the flow so that the inner morphological zone acts as a normal marsh creek network, with low-energy input rapidly dissipated in the network of narrow creeks.

North Norfolk marshes Figure 3.8 shows a marsh creek network on Scolt Head Island in North Norfolk. This branched creek system, it is suggested, is a response to the energy regime within the outer tidal channel known as Norton Creek. This channel is one which is open at both ends and, therefore, does not exhibit the resonant characteristics of a closed estuarine channel. This means that a medium-magnitude tidal energy input affects the saltmarshes fed by Norton Creek which, moreover, since it lies in the shelter of the Scolt Head Island barrier, experiences very little wind-generated wave energy. As a consequence, the tidal energy here can be rapidly dissipated as tidal flow enters, and is sub-divided within, the creek network. The repeated bifurcation of the creeks, together with the exponential rate of decrease in total creek width, means that individual creeks become extremely narrow towards the extremity of the network. This means that tidal energy is rapidly dissipated and, by the time the flood tide reaches the first order channels of the network, current velocities are negligible. These first order channels are so narrow as to prevent further bifurcation taking place since, if they were to do so, the decrease in width would mean the total closure of the channel. Thus creek development is halted at some distance from the back-marsh edge, both by the reduction in tidal velocity and the morphological restrictions on further creek bifurcation as may be seen in Fig. 3.8.

Blackwater Estuary Marshes Finally, the marsh creek networks of the Blackwater estuary may be described. Figure 3.9 shows a section of the Tollesbury marshes. Here the estuary tides in a resonant channel introduce high-energy flows into the mudflat-saltmarsh system. The marsh creek network responds to these energy inputs, in the manner described in the last section, by adopting a branched network in which creek width narrows exponentially away from the creek mouth. In the Tollesbury marshes, however, tidal energy would be too great to allow total energy dissipation within the confines of a narrow marsh if bifurcation forced channel length limitation in the same way as in the North Norfolk case. Instead, the creek system here develops a series of intricate meanders which lengthen the

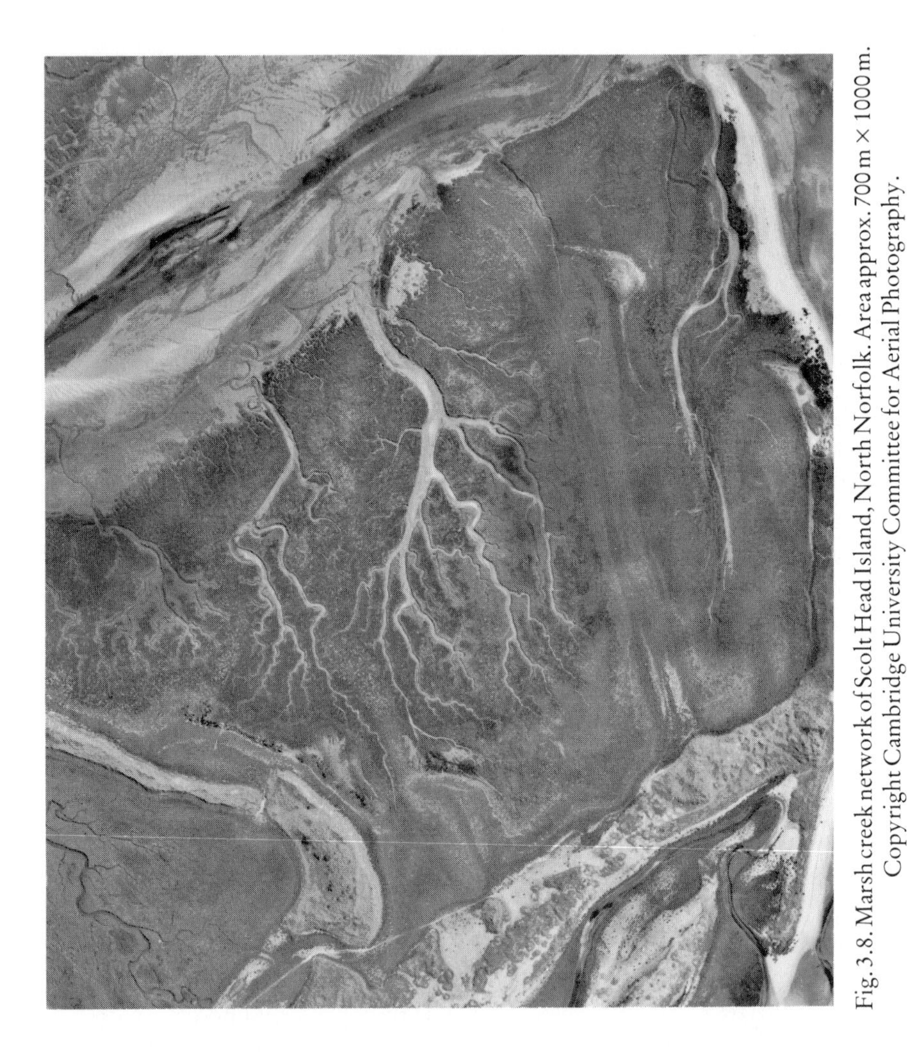

Fig. 3.8. Marsh creek network of Scolt Head Island, North Norfolk. Area approx. 700 m × 1000 m. Copyright Cambridge University Committee for Aerial Photography.

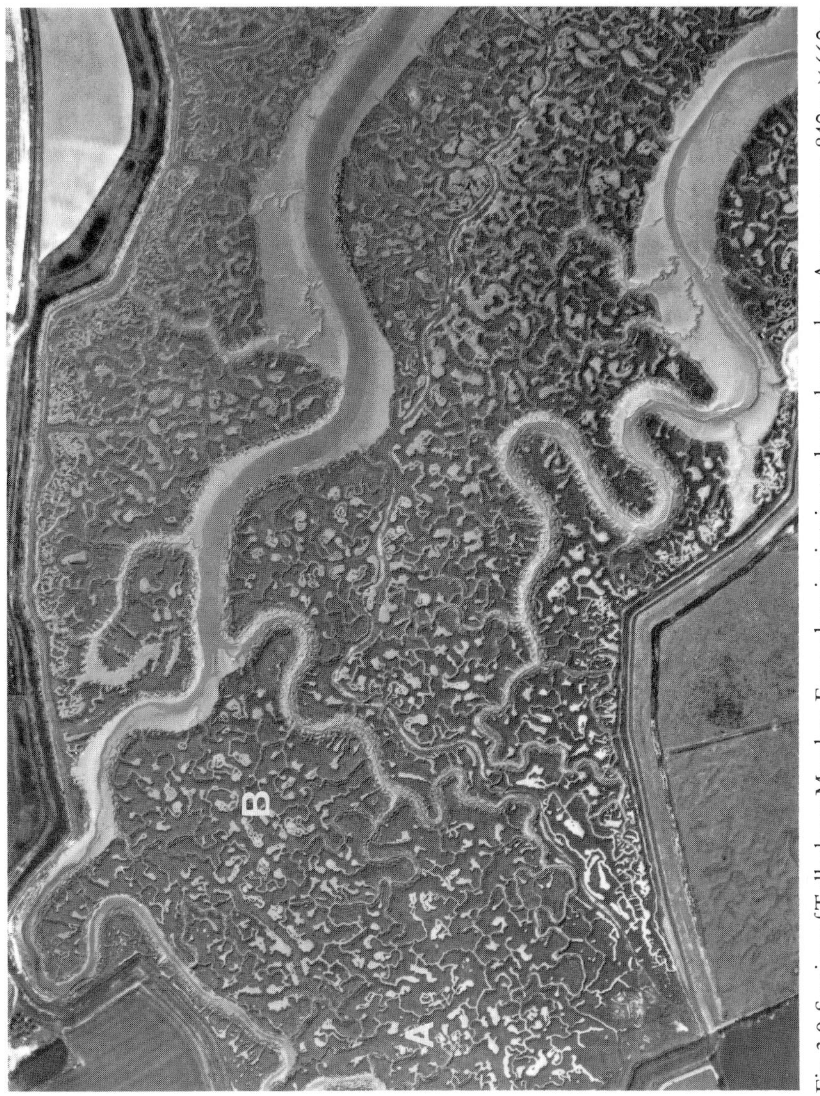

Fig. 3.9. Section of Tollesbury Marshes, Essex showing intricate channel meanders. Area approx. 840 m × 660 m. Copyright Cambridge University Committee for Aerial Photography.

channels without bifurcation. This forces the tidal flow through longer channels than would be possible if the network were to consist of relatively straight sections with frequent bifurcations, a morphological device which allows more frictional retardation of the tidal wave on a narrow marsh.

Sea-level rise and saltmarsh morphology

The effects of sea-level rise on saltmarsh morphology have already been mentioned in this paper. Increase in water levels means a decrease in return intervals for given wave magnitudes and this, in turn, means that the marsh recovery from wave events may not be sufficiently rapid to prevent permanent recession of both the edge and the surface. In fact, measurements of surface elevation in Essex indicate that recovery of the surface elevation is rapid enough, at present, to prevent long-term degradation by the increasing frequency of erosional events.

The same is not true of the saltmarsh edge on the Essex marshes. Here, long-term recession has resulted from the increasing frequency of storm events, without sufficient edge recovery in the intervening phases. The long-term effect, therefore, is that the mudflats of this part of the coast are progressively flattening and increasing in length at the expense of the saltmarsh. This is, as described above, a classic coastal response to increased wave energy, although in this case it is a long-term trend rather than a periodic adjustment. An important aspect for coastal management here is that the erosion of the saltmarsh does represent a natural response to sea-level rise; any attempts to protect the saltmarsh edge, therefore, by wave breaks or other constructions, will mean that the mudflat section of the intertidal zone will be prevented from adjusting to the new energy conditions, a process which will be exacerbated by the loss of released sediment from saltmarsh erosion.

The response of the estuarine creeks to sea-level rise is also one which represents a long-term trend towards increased energy dissipation, but in this case for tidal energy. The creek system on the Essex estuarine saltmarshes has developed as a response to a given tidal prism and energy input. As the sea level rises so both tidal prism and tidal energy increases in the creeks, so that, as the tidal flow moves landwards, not all the energy can be dissipated. The result is that as the flood tide reaches the landward end of the creek a significant flow velocity is still present, sufficient to cause undermining of the creek banks. The creek cross section at the landward end is characterised by an extremely low width/depth ratio: 20 cm wide and 1 m deep is common. The effect of bank erosion is to cause the high, vertical banks to collapse, so widening the creeks at their landward end. The result is the development of a series of near-circular pools at the end of each creek (e.g. at point A, shown in Fig. 3.9), which thus lengthen the creek sufficiently to accommodate the new energy regime. In some cases, as on the Tollesbury marshes, creek lengthening has proceeded by the development of a concentric whorl in the landward end of the creek system (e.g. at point B, Fig. 3.9), a morphological response the effect of which is explained by the above discussion but whose formative process is not so obvious.

Conclusion

The discussion in this contribution has attempted to provide a unified framework for geomorphology of saltmarshes. In so doing it has necessarily posed more problems than it has answered, and it is clear that research work is urgently needed if any attempt is to be made to understand our low-energy coastlines before they disappear as sea level rises. Saltmarshes are seen here as the safety valve of the intertidal zone; erosion of their surface and edges is a natural response to imposed energy conditions. The fact that saltmarshes exist is proof enough that they have, in the past, recovered from such erosional events. The overall response of the intertidal profile to high-magnitude events, whether they are tidal- or wind-generated, is for them to increase in to length so as to extend the wave-sediment interface. This simple adjustment leads, in detail, to some extremely complex and varied morphological solutions, and it has been the aim of this paper to place some of these complexities back into their simple framework. Sea-level rise in southeast England has changed the frequency of these morphological adjustments, so that, without changing any of the processes involved, a long-term geomorphological trend has resulted. Such a development may be expected to affect many of the marshes, and indeed mangroves, of the world if the predicted sea-level rise due to global warming takes place. The long-term outlook for saltmarshes, therefore, especially in those areas where sea walls prevent landward transgressions, must be viewed pessimistically. Yet the role of the saltmarsh is one of a long-term insurance policy to be used in times of emergency; the present state of saltmarshes is merely the fulfilment of that role and it may be that the sacrifice of the saltmarshes will mean the eventual geomorphic adjustment of the low-energy coastal profile to sea-level rise.

References

ALLEN, J.R.L. (1989) Evolution of salt-marsh cliffs in muddy and sandy systems: a qualitative comparison of British west-coast estuaries. *Earth Surface Processes and Landforms* **14**, 85–92.

ALLEN, J.R.L (1990) Salt-marsh growth and stratification: a numerical model with special reference to the Severn Estuary, south-west Britain. *Marine Geology* **95**, 77–96.

ANDERSON, F.E. (1972) Resuspension of estuarine sediments by small amplitude waves. *Journal of Sedimentary Petrology* **42**, 602–607.

CARTER, R.W.G. (1988) *Coastal Environments.* Academic Press, London.

FRENCH, J.R., SPENCER, T., & STODDART, D.R. (1990) Backbarrier Marshes of the North Norfolk Coast: Geomorphic Development and Response to Rising Sea Levels. *University College London, Discussion Papers in Conservation* **54**, 28 pp.

LANGBEIN, W.B. (1963) The hydraulic geometry of a shallow estuary. *International Association of Scientific Hydrologists* **8**, 84–94.

LEOPOLD, L.B., WOLMAN, M.G., & MILLER, J.P. (1964) *Fluvial Processes in Geomorphology.* Freeman, San Francisco.

MYRICK R.M. & LEOPOLD L.B. (1963) The geomorphology of a small tidal estuary. *US Geological Survey Professional Paper* **422B**, 1–18.

PETHICK, J.S. (1978) Further notes on the drainage density-basin area relationship. *Area* **10**, 249–50.

PETHICK, J.S. (1981) Long-term accretion rates on tidal saltmarshes. *Journal of Sedimentary Petrology* **51**, 571–577.

PETHICK, J.S. (1991) *Essex Saltmarsh Erosion* (7). Report to National Rivers Authority, Anglian Region.

PETHICK, J.S. (In press) Saltmarsh creek morphology. In D.R. Stoddart (ed.), *Saltmarshes and Coastal Wetlands.* Blackwell, Oxford.

REED, D.J. (1988) Patterns of sediment deposition in subsiding coastal marshes, Terrbone Bay, Louisiana: the role of winter storms. *Estuaries* **12**, 222–227.

SHEPARD, F.P. & LAFOND, E.C. (1940) Sand movements near the beach in relation to tides and waves. *American Journal of Science* **238**, 272–285.

STUMPF, R.P. (1983) The processes of sedimentation on the surface of a saltmarsh. *Estuarine and Coastal Marine Science* **17**, 495–508.

WRIGHT, L.D., COLEMAN, J.M. & THOM, B.G. (1973) Processes of channel development in a high-tide range environment: Cambridge Gulf-Ord River Delta, Western Australia. *Journal of Geology* **81**, 15–41.

4

Saltmarsh plant ecology: zonation and succession revisited

A.J. GRAY

Introduction

This contribution highlights a series of current research issues in the plant ecology of saltmarshes. In choosing which issues to include I have largely avoided the role of vegetation in sediment processes and dynamics. This is not because of its lack of importance – on the contrary, it is presently a central and exciting topic – but because it is likely to be covered in other contributions by authors more qualified than I to address it. However, towards the end of the paper I include some of our recent work on sedimentary changes in different plant communities.

In selecting issues, it has been illuminating to return to two problems which occupied the early saltmarsh ecologists (typified by V.J. Chapman), namely, the causes of zonation and succession, and to discover that not only is there much currency in such topics, there is also a great deal we have yet to learn about them.

Zonation: the niche and competition

From a plant ecologist's viewpoint saltmarshes are remarkable in at least two major respects. First, they comprise large areas of natural vegetation undergoing primary succession and containing relatively few plant species. They thus provide an opportunity to examine the conditions under which communities are assembled *de novo* largely without human interference. Secondly, saltmarshes frequently display strong spatial structure, usually in the form of parallel zones of vegetation, the communities changing with increasing surface elevation. Again the opportunity to investigate the processes, which in some marshes may produce strikingly sharp demarcation lines, is manifest. Despite these opportunities, studies of community assembly, spatial structure and succession are more often made in multi-

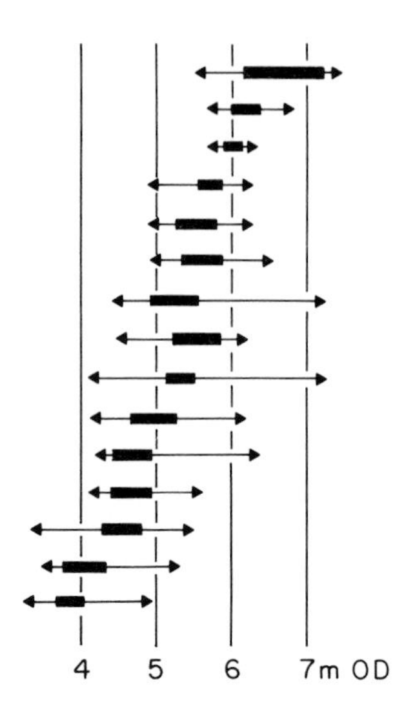

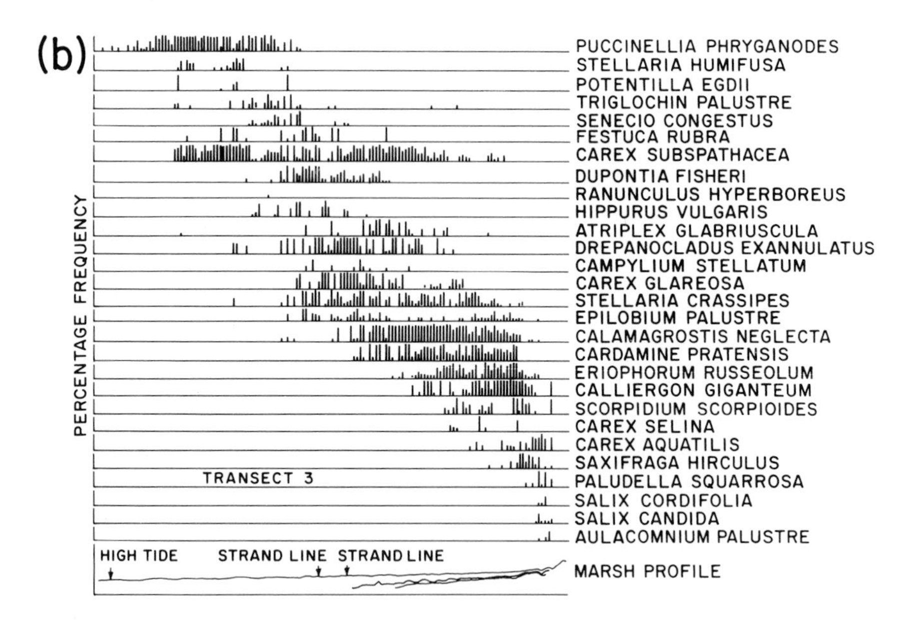

Fig. 4.1. Zonation patterns in species distributions on saltmarshes in (a) Morecambe Bay, UK (above) and (b) Hudson Bay (Canada) (below after Kershaw 1974). The arrows in the top diagram illustrate total range; the bars the main range.

species, spatially complex and anthropogenic habitat types such as chalk grassland. Whether by perversity or mistrust of the simple, ecologists have consistently eschewed a chance to 'explain' plant community structure.

The niche Zonation in saltmarshes is generated by the different, but often overlapping, vertical ranges of individual species. Two examples given in Fig. 4.1, one from a British saltmarsh system in Morecambe Bay and one from Hudson Bay in Canada, are typical of the results obtained from analyses of species' distributions along transects from low to high water. Individual species' ranges are thought to reflect their relative ability to tolerate tidal submergence, or at least factors related to tidal submergence such as soil anaerobis, under conditions where there is also increasing competition with other species as elevation increases (Gray, 1980, 1985). The general dogma is that the lower vertical limit is controlled largely by tolerance of tidal factors and the upper limit is fixed by interspecific competition.

This is, of course, far too simple. However, a useful concept to apply here is Hutchinson's view of the 'fundamental' versus the 'realised' niche (Hutchinson, 1957). The fundamental niche is the total biological and environmental space (n-dimensional hypervolume) which a species may potentially occupy, as opposed to the realised niche which comprises the space species actually occupy, occupancy of parts of their fundamental niche being prevented usually by interspecific competition or lack of opportunity to spread there. In the context of saltmarsh plants, the gradient of tide-related factors which change with increasing elevation can be viewed as one dimension of the niche. In the absence of other species, any one species will occupy a section of that gradient which can be defined as its fundamental niche. As the number and type of other species are added, so the ability of the species of interest to occupy its fundamental niche is likely to diminish. The idea that as species number and diversity increase, the increase in competitive interactions narrows the niche (or the utilisation of a resource), termed 'diffuse competition' by MacArthur (1972), can be tested in the field.

For example, Pielou and Routledge (1976) showed, as predicted, that the zones occupied by some saltmarsh species on North American marshes became wider as species diversity decreased. More recently, in a detailed study, Russell *et al.* (1985) demonstrated a similar pattern in two English south-coast marshes only 6 km apart, the niche breadths of eight out of nine species common to both marshes being broader on the marsh with the lower species number (11 versus 16) and diversity (a Shannon-Wiener index of 1.59 versus 2.28). Here niche breadth was defined after Levins (1968), the resource being 'height of marsh above tide datum'. The Russell *et al.* results also agreed with Pianka's niche overlap hypothesis that niche overlap should decrease with increasing intensity of competition (Pianka, 1974). The mean niche overlap of species on the marsh with the highest species number and diversity was significantly lower than on the other marsh.

The effect of increasing diffuse competition on niche overlap and breadth is a major reason why it is difficult to quantify the niche of a species across a range of saltmarshes varying in floristic composition and diversity. This is particularly relevant at higher elevations where more species occur, and hence more combinations of biological interactions are possible. However, a recent study of *Spartina anglica* suggests that attempts to quantify the niche of species in the lower zones, where there are fewer species and a predominance of physical limiting factors, may be a fruitful line of future research. In the *Spartina* study, Gray *et al.* (Gray, Clarke, Warman and Johnson, 1989 and unpublished; Gray, Marshall and Raybould, in press) compared the vertical range of *Spartina anglica* along 107 line transects across saltmarshes in 19 estuaries in south and west Britain. A total of 27 physical variables was measured for each transect, including measurements of tide range, marsh size, slope and exposure, estuarine area and aspect, and latitude and sediment type. Multiple regression was used to examine the relationship between the physical parameters and the elevational limits of *Spartina*. Although, remarkably, variation in tidal range was able to account for between 86 and 89% of the variation in limits, the addition of other variables did improve their prediction. The equation:

$$LL = -0.805 + 0.366\,SR + 0.053\,F + 0.135\,\mathrm{Log}_e A$$

where LL = lower limit of *Spartina* (in m AOD), SR = spring tidal range (m), F = fetch in the direction of the transect (km) and $\mathrm{Log}_e A$ = $\log_e$ estuary area (km^2), had an r^2 of 93.7 and s (residual standard deviation in metres) of 0.35. It seems, therefore, that *Spartina* extends further downshore than would be predicted from the effects of tidal range alone on transects with a shorter seaward fetch and in smaller estuaries (possibly because of the effects of exposure and of wind-generated waves at high tide).

The upper limit of *Spartina* was described by:

$$UL = 4.74 + 0.483\,SR + 0.068\,F - 0.199\,L$$

where SR and F are as above and L = latitude (in degrees N expressed as a decimal). Here $r^2 = 90.2$ and $s = 0.50$. Clearly the upper limit (very often where *Spartina* adjoined a *Puccinellia maritima*-dominated community) is similarly affected by fetch but also varies significantly with latitude when the effects of tidal range and fetch have been accounted for, the further north the marsh the lower down the shore is the upper limit of the *Spartina* sward. This may well be related to the changing competitive interaction with *Puccinellia* and other species under northern climatic conditions; *Spartina* is one of only eight species in the British Flora that utilise the C_4 photosynthetic pathway and mainly occur in tropical and subtropical regions, and the growth of which is more greatly influenced by temperature (Long *et al.*, 1975; Long, 1983).

Niche models of the type described for *Spartina* have at least two important functions. First, they provide a means of predicting the extent to which a species may occupy or invade an area. For example, the potential area of mudflat which *Spartina* might occupy in an estuary it has not yet invaded or where it has not yet expanded can be predicted from a simple levelling survey of the mudflats and data on tidal range, fetch, estuary area, and latitude. That we can account for more than 90% of the variation in elevational limits of *Spartina* is remarkable and gives confidence to such predictions. It also means that predictions can be made for the distribution of *Spartina*, and potentially other saltmarsh species, following alterations in tidal range due to structures such as tidal power barrages in estuaries. (There are complications in such predictions relating to the changes in tidal regime which follow barrages which cannot be expanded upon here but are partly covered in Gray *et al.* 1989.)

A second function of such niche models is to generate questions and hypotheses about the processes controlling population and community structure. Why, for example, is it possible to account for such a remarkably high proportion of the variation in *Spartina* limits? One possibility, alluded to above, is that the precision reflects the fact that the species, occupying the lowest part of the intertidal zone throughout most of its range, is limited in its spread by physical, more than biological, factors. Such a hypothesis is supported by work on the species' tolerance of tidal immersion and on the physical disturbance of seedlings at different mudflat levels (Hubbard, 1969;

van Eerdt, 1985; Groenendijk, 1986). Further, it is a testable hypothesis which demands transplant and other experiments of the type advocated by Ranwell (1972) to compare dispersal and establishment in the adult and seedling stages.

Competition Niche theory also provides a theoretical framework within which to explore plant species' interactions. For instance, the co-occurrence of species along the altitudinal gradient cannot be regarded as niche overlap unless competition is occurring between them. They may be coexisting, sharing a resource that exceeds their combined demands, or utilising different resources (have a different niche dimension) within the same area (Colwell and Futuyma, 1971). One way of testing for species interactions in the field is to remove one or more species from a community to see how the growth of other species is affected (Silander and Antonovics, 1982; Fowler, 1984).

Such an experiment was performed in The Netherlands by Scholten and Rozema (1990) in the transition zone between *Spartina* and *Puccinellia* communities. Either *Spartina* or *Puccinellia* was removed from sections of two adjacent plots differing in elevation by about 4 cm, corresponding to a 15–20 minute difference in tidal submergence (but with no major differences in salinity, soil texture or redox potential). The results, depicted in Fig. 4.2, are extremely revealing. Removal of *Spartina* in the lower plot caused a significant increase in Puccinellia biomass production (i.e. significantly higher than in mixed plots where neither species had been removed) but its removal from the higher plot did not. Conversely, removal of Puccinellia in the higher plot caused a significant increase in *Spartina* production, an effect not found in the lower plot. This suggests that the competitive interaction between these species within the overlap zone is asymmetric, *Spartina* being depressed by Puccinellia in one area, the higher plot, and the reverse occurring in another area, the lower plot. The competitive advantage appears to have changed over a very short distance and small rise in elevation.

Further experiments are needed to explore the mechanisms of competition between *Spartina* and *Puccinellia*. Scholten and Rozema (1990) themselves report several others, but the findings of such simple removal experiments can be used to plan the management of saltmarshes. They suggest, for example, that repeated clipping or grazing can be used to transform many *Spartina*-dominated marshes

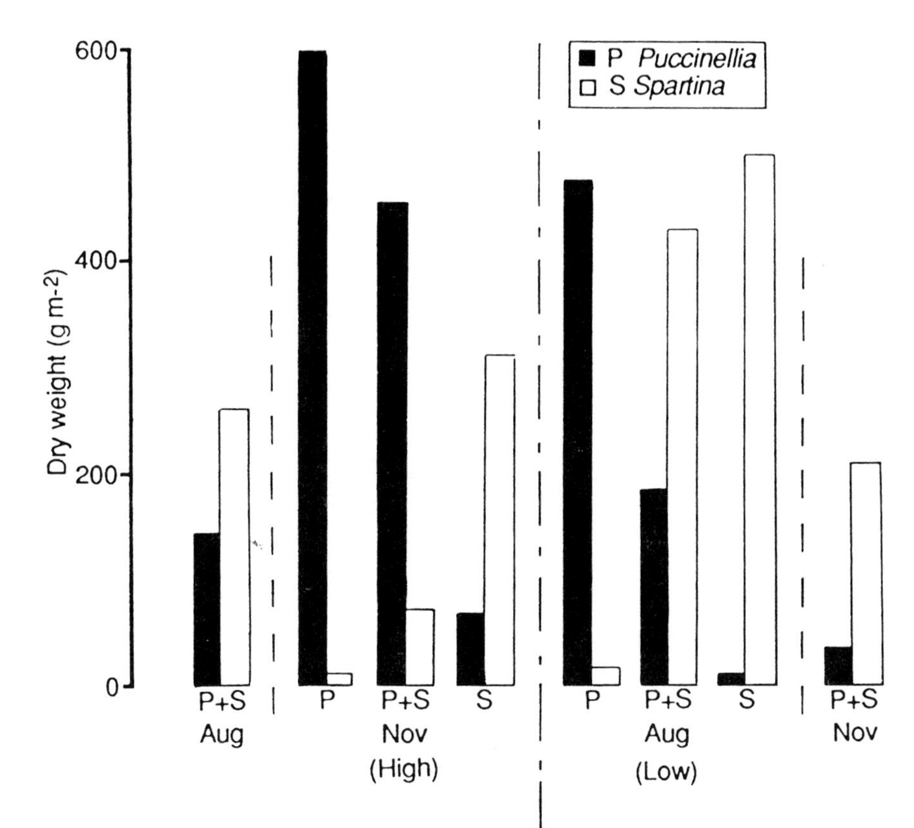

Fig. 4.2. Shoot biomass of *Puccinnellia* and *Spartina* at the start (August) and end (November) of a deletion experiment in both higher (high) and lower (low) plots. P+S, no removal, P = *Puccinellia* (monoculture) (*Spartina* removed), S = *Spartina* monoculture (*Puccinellia* removed). After Scholten & Rozema (1990).

to *Puccinellia* marsh, a suggestion confirmed by the *Spartina* harvesting experiments of Scott *et al.* (1990).

The importance of marsh surface elevation in determining the outcome of competitive interactions was also revealed in three-species competition experiments involving *Puccinellia maritima*, *Festuca rubra* and *Agrostis stolonifera* (Gray and Scott, 1977). On the grazed saltmarshes of Morecambe Bay, these species coexist in mosaic populations at the higher marsh levels, a survey of microtopographical variation in their zone of overlap showing that *Puccinellia* is largely restricted to hollows, *Festuca* to humps, and *Agrostis* to the edges of humps or in hollows at higher elevations than *Puccinellia*

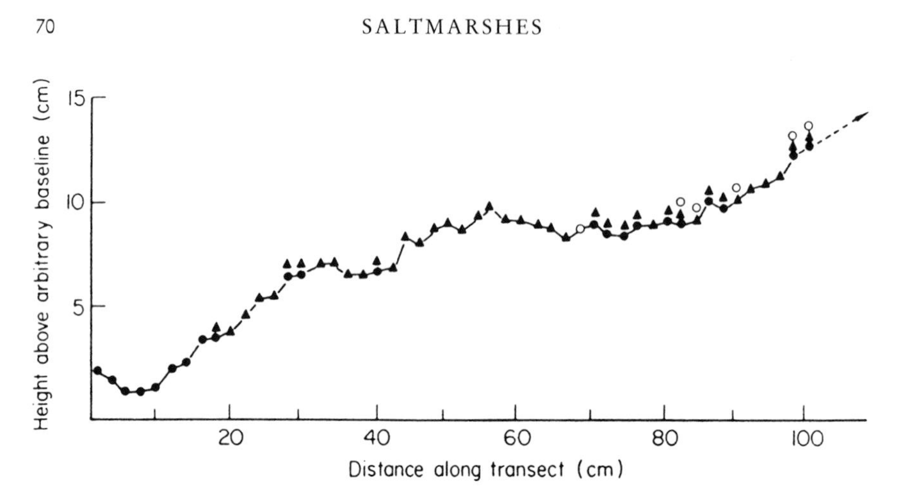

Fig. 4.3. The distribution of *Puccinellia* (●), *Festuca* (▲) and *Agrostis* (○) in relation to microtopography along a transect on Silverdale Marsh. The symbols indicate the presence of each species, determined by contact with a pin 3 mm in diameter. After Gray & Scott (1977).

(Fig. 4.3). Replacement series pot competition experiments (de Wit, 1960) in which salinity, water level and soil type were varied, indicated that these small-scale distribution patterns are determined by the species' relative competitive abilities under various conditions of waterlogging and salinity. For example, the competitive interaction between *Puccinellia* and *Festuca*, both fairly salt-tolerant species, was strongly affected by waterlogging (Fig. 4.4) so that *Festuca* outyielded *Puccinellia* in 2:2 mixtures in the driest conditions but was outyielded by *Puccinellia*, even in 3:1 mixtures, at the highest water level. *Agrostis* was strongly competitive in this experiment, generally having convex curves with both companion species (Fig. 4.4), but in other experiments proved to be a poor competitor under saline conditions (Gray and Scott, 1977). Therefore, as has been demonstrated by other studies (e.g. Brereton, 1971), subtle microtopographical differences can determine and maintain species distributions in their zones of overlap.

Succession and community dynamics

It is necessary, for completeness, to reiterate a cautionary note about the relationship between zonation and plant succession on saltmarshes (although briefly, since the subject has been covered elsewhere, particularly in relation to population differentiation and

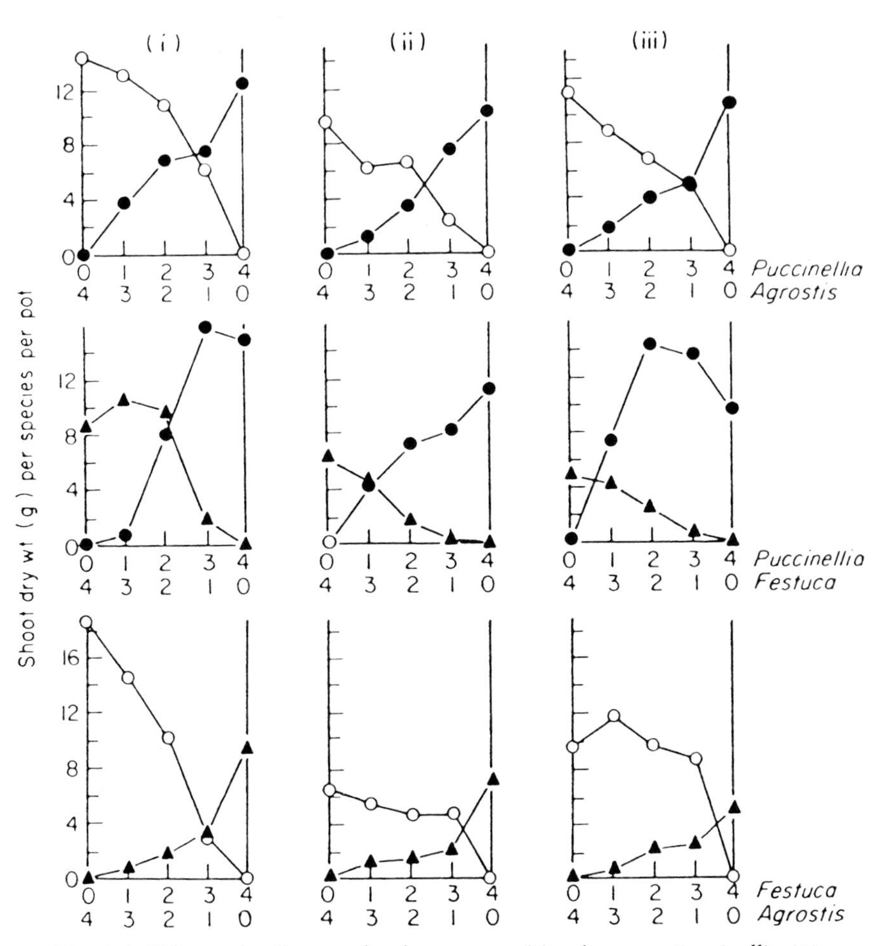

Fig. 4.4. Effects of soil water level on competition between *Puccinellia* (●), *Festuca* (▲) and *Agrostis* (○) grown at various ratios. (i) Water level at pot base; (ii) water level 5 cm below soil surface; (iii) water level at surface. After Gray & Scott (1977).

genetic structure: Gray *et al*., 1979; Gray, 1985, 1987). Essentially, it must not be assumed that zonation along an elevational gradient (a spatial feature) has been wholly, or even partly, generated by succession (a temporal phenomenon). Demonstration of competitive interaction at a zone boundary, or niche overlap, does not indicate that one plant community has developed from that below it. It may, as in the examples above, indicate the mechanism of competitive displacement and even the major factors (e.g. elevation-related variables) conferring competitive advantage, but it cannot prove that the process is dynamic, that the vegetation in the upper zones has developed from a type currently characterising the lower zones. The relatively recent colonisation of many of the lower zones of existing saltmarshes by *Spartina* is an excellent illustration of the temporal disjunction between adjacent vegetation types.

Our knowledge that saltmarsh succession is widespread, and that plant communities commonly displace those found in the zone immediately below them, stems from (a remarkably few) long-term monitoring studies of permanent quadrats and from perturbation experiments of the type described earlier. Although changes in surface elevation over time, resulting from sediment accretion within saltmarshes, have been measured in several studies, and despite the obvious importance of elevation as an agent of community change, there have been few attempts to produce dynamic quantitative models of community change related to elevation (or indeed, as Pethick *et al.* (1990) point out, to understand the physical processes which result in deposition on a vegetated surface). An exception is the study of Randerson (1979) (and later unpublished work by M. Hill and P.F. Randerson, pers. comm.) on the saltmarshes of The Wash.

This is a fairly complicated model (Fig. 4.5), but its development highlighted some previously unsuspected relationships – for example, that between sediment accretion rates and below-ground, rather than above-ground, plant biomass. In addition, the behaviour of the model in simulations enabled adjustments to accretion rate parameters (specifically to the threshold level at which they were switched between upper and lower marshes) to provide a more realistic time frame. Such insights into both qualitative and quantitative relationships between plant community and sedimentary changes are extremely valuable, providing both a basis on which to seek further empirical data and more understanding of the dynamics of the system. Other work on the same marshes, at the same time, by Coles (1979) established the importance of benthic microalgae in mudflat stabilisation and suggested that the processes of biogenic accretion should somehow be incorporated into mudflat/saltmarsh sedimentation models.

Having finally arrived at the point where the plant ecologist's interests closely coincide with those of the geomorphologist (at the zone of niche overlap?) I will describe briefly the preliminary results of a study, which is still in progress, of sediment changes in two south coast saltmarshes (A.J. Gray, E.A. Warman and P.E.M. McKinlay, unpublished). As part of this study the patterns of sediment change

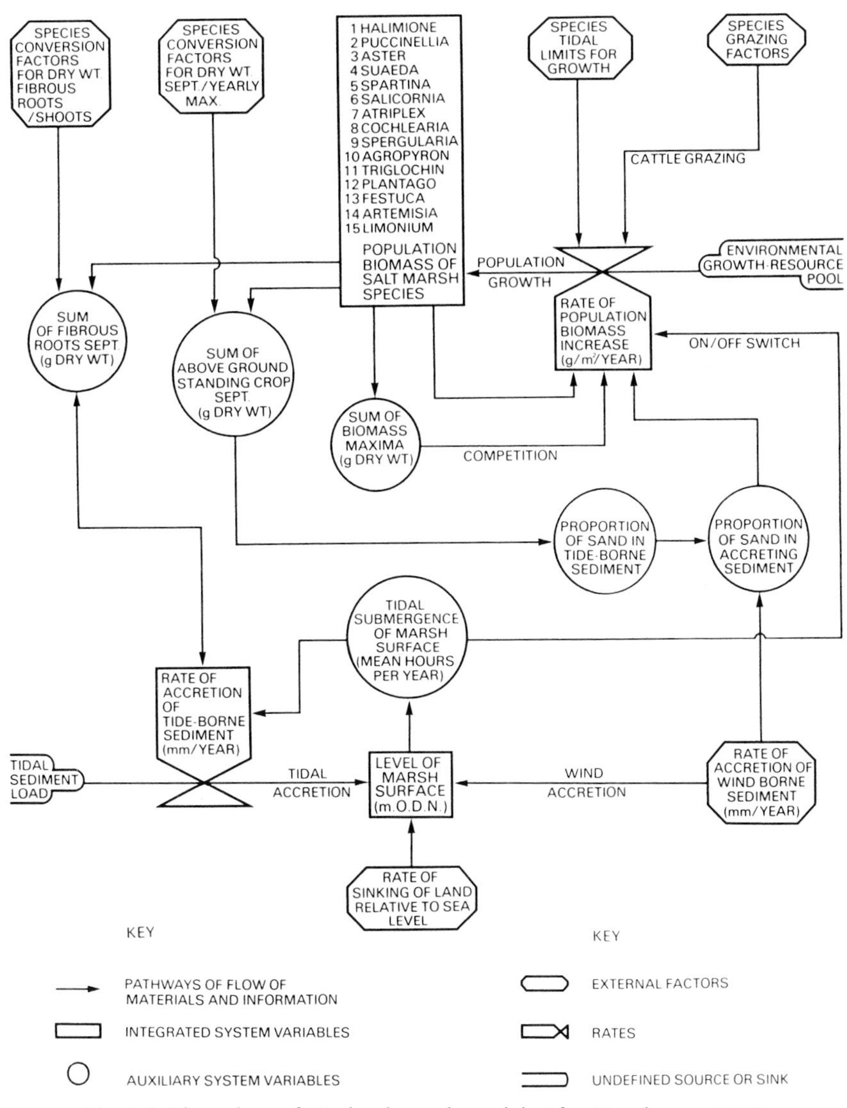

Fig. 4.5. Flow chart of Wash saltmarsh model. After Randerson (1979).

were measured in similar vegetation types on two contrasting saltmarshes around Furzey Island in Poole Harbour, Dorset. The marshes display similar vegetation zones with, from seaward to landward, an area of open mudflat with scattered *Spartina* clumps giving way to a continuous *Spartina* sward (containing *Salicornia perennis*), a zone dominated by *Halimione portulacoides* (with *Puccinellia maritima* common), and a 'species-rich' zone usually with

Table 4.1. *Mean elevation and sediment erosion/accretion rates as measured by molochite cores in comparable vegetation types on two saltmarshes in Poole Harbour. Elevation is in m above OD. Accretion/erosion is in mm a^{-1}. The range of values is given in brackets.*

Vegetation zone		NE Marsh	S Marsh
Mudflats	(Elevation)	0.41	0.39
	(Accretion)	+34.0	+3.90 (−20.0 to +75.3)
Spartina	Elevation	0.49 (0.19–0.65)	0.45 (0.20–0.52)
	Accretion	+6.50 (+<1.0 to +8.9)	+1.10 (−13.4 to +11.7)
Halimione	Elevation	0.54 (0.44–0.64)	0.71 (0.55–0.92)
	Accretion	+6.10 (+1.5 to +9.3)	+2.90 (+1.0 to +5.3)
Species rich	Elevation	0.59 (0.47–0.66)	0.62 (0.52–0.72)
	Accretion	+1.70 (+<1.0 to +4.5)	+3.30 (+1.0 to +5.3)

Plantago maritima, *Triglochin maritima*, *Armeria maritima*, *Limonium vulgare*, *Puccinellia maritima* and *Aster tripolium* and having *Festuca rubra* and *Juncus gerardii* at its upper edge. At higher elevations a fringe of *Elymus pycnanthus*, with other drift-line species such as *Beta maritima* and *Atriplex hastata*, marks the upper limit of the intertidal marsh. Although similar in their plant communities, the marshes have a different history and sedimentary environment. One, on the northeast of the island, has developed behind a sand and gravel bar, derived from local erosion of Bagshot deposits, and is flooded via a single channel in one corner, the tide overspilling a dendritic creek system onto the marsh surface. This marsh (NE marsh) appears to have been gradually developing over the last 100 years. Its sheltered position and predominantly accretionary environment is in contrast to the marsh on the south of the island (S marsh) which, like nearly all of the saltmarshes in Poole Harbour, has experienced extensive dieback of *Spartina* along its lower edge over the past 60 years (Gray and Pearson, 1984). The marsh is directly flooded along its entire front by the rising tide.

A series of line transects (20 in all) across both marshes was levelled to Ordnance Datum and marker cores were placed at intervals along these. The cores comprised a T-shaped plug of molochite (a chemically-inert calcined china clay supplied to us by English China Clays) 25 mm in diameter and 75 mm deep, which were placed in batches of three and recovered after periods of 18 months to 3 years

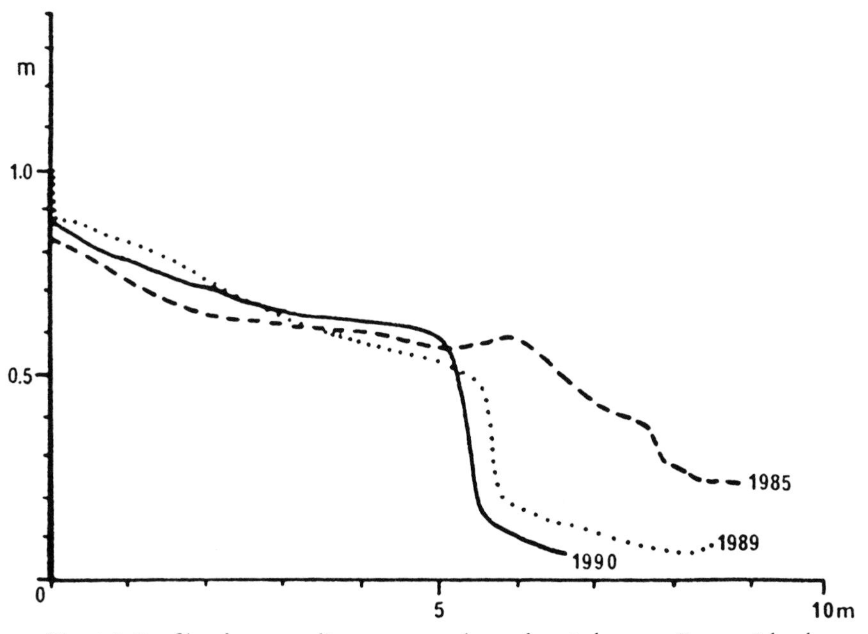

Fig. 4.6. Profile of transect line across a saltmarsh at 3 dates on Furzey Island, Poole Harbour (see text for explanation).

(187 cores so far removed). The annual changes in erosion/accretion rates within similar plant communities on the two marshes are given in Table 4.1. On the NE marsh, accretion, ranging from a trace to 34 mm a^{-1}, occurred over all cores in all zones, with a decline in accretion rate with height and marsh maturity, as might be predicted from studies such, as those of Letsch and Frey (1980) and Pethick (1981). In contrast on the S marsh many cores were eroded in the mudflat and *Spartina* zones, the rather small average net accretion rates being misleading in view of the high variance. Above these extremely variable zones, the *Halimione* and species-rich zones did show continuous accretion. In fact, comparison of the pattern of accretion at the higher levels on the two marshes reveals that, although it was lower (2.9 versus 6.1 mm a^{-1}) in the *Halimione* zone (which it is important to note develops at a lower elevation on the more sheltered NE marsh) on the S marsh, the accretion rate in the species-rich zone on this marsh was around twice that on the NE marsh. The marsh from which *Spartina* is being eroded thus displays relatively higher (enhanced?) accretion rates at higher elevations. If, as seems likely, this is related to the increased mobilisation of sediments in the lower zones, the steepening of the marsh profile and

cliff formation appear to be an inevitable consequence. Such a process can be seen from a comparison of transect line profiles at three dates on a third marsh on Furzey Island (Fig. 4.6). This marsh is also fronted by extensively eroded mudflats formerly carrying *Spartina* marsh. The landward advance, steepening and increasing height of the cliff is accompanied by the deposition of sediment in the upper marsh.

The research described above has yet to be completed but the implications of the preliminary findings for predicting the changes in saltmarsh plant communities as *Spartina* marshes disappear along the south coast are clear. Moreover, it is likely that cliff formation, the loss of lower zones and, by accelerated succession, the invasion of the middle zones by high marsh species, may produce a very different set of changes in these marshes of small tidal range from those predicted elsewhere under rising relative sea levels (Boorman *et al.*, 1989). The modelling of such changes is an obvious and challenging prospect.

Concluding remarks

In selecting some current issues around the subjects of zonation and succession I have omitted several important recent developments in saltmarsh plant ecology. This is particularly true for the fields of plant production, decomposition and nutrient cycling where, for example, in confronting the problem of predicting the effects of possible global climate warming, we are now armed with Long's (1990) elegant model of the very different responses of *Puccinellia* and *Spartina*. Nor have studies of saltmarsh management, notably of grazing (partly covered elsewhere by Gray, 1991), been included. Nonetheless, it is hoped that the selection of problems discussed above is sufficient to appraise the ecologist of the wealth and potential of research on a relatively simple system, and to disabuse the geomorphologist of the view that saltmarsh vegetation is merely the icing on a cake fashioned by physical processes.

References

BOORMAN, L.A., GOSS-CUSTARD, J.D. & McGRORTY, S. (1989) *Climatic Change, Rising Sea-level and the British Coast.* HMSO, London, 24 pp.

BRERETON, A.J. (1971) The structure of the species populations in the initial stages of saltmarsh succession. *Journal of Ecology* **59**, 321–338.

COLES, S.M. (1979) Benthic microalgal populations on intertidal sediments and their role as precursors to saltmarsh development. In R.L. Jefferies & A.J. Davy (eds) *Ecological Processes in Coastal Environments.* Blackwell Scientific Publications, Oxford, 25–42.

COLWELL, R.K. & FUTUYMA, D.J. (1971) On the measurement of niche breadth and overlap. *Ecology* **52**, 567–576.

FOWLER, N. (1984) Competition and coexistence in a N. Carolina grassland. II. The effects of the experimental removal of species. *Journal of Ecology* **69**, 843–854.

GRAY, A.J. (1980) Saltmarshes and reclaimed land. In G. Halliday & A.W. Malloch (eds) *Wild Flowers: Their Habitats in Britain and Northern Europe.* Peter Lowe, London, 123–133.

GRAY, A.J. (1985) Adaptation in perennial coastal plants – with particular reference to heritable variation in *Puccinellia maritima and Ammophila arenaria. Vegetatio* **61**, 179–188.

GRAY, A.J. (1987) Genetic change during succession in plants. In A.J. Gray, M.J. Crawley & P.J. Edwards (eds) *Colonization, Succession and Stability.* Blackwell Scientific Publications, Oxford, 273–293.

GRAY, A.J. (1991) Management of coastal communities. In I.F. Spellerberg, M.G. Morris & F.B. Goldsmith (eds) *Scientific Management of Temperate Communities for Conservation.* Blackwell Scientific Publications, Oxford, 227–243.

GRAY, A.J. & PEARSON, J.M. (1984) *Spartina* marshes in Poole Harbour, Dorset, with particular reference to Holes Bay. In J.P. Doody (ed.) *Spartina anglica in Great Britain.* Nature Conservancy Council, Attingham, 11–14.

GRAY, A.J. & SCOTT, R. (1977) The ecology of Morecambe Bay. VII. The distribution of *Puccinellia maritima, Festuca rubra* and *Agrostis stolonifera* in the saltmarshes. *Journal of Applied Ecology* **14**, 229–241.

GRAY, A.J., PARSELL, R.J. & SCOTT, R. (1979) The genetic structure of plant populations in relation to the development of saltmarshes. In R.L. Jefferies & A.J. Davy (eds) *Ecological Processes in Coastal Environments.* Blackwell Scientific Publications, Oxford, 43–64.

GRAY, A.J., CLARKE, R.T., WARMAN, E.A. & JOHNSON, P.J. (1989) *Prediction of Marginal Vegetation in a Post-barrage Environment.* Energy Technology Support Unit Report ETSU-TID-4070.

GRAY, A.J., MARSHALL, D.F. & RAYBOULD, A.F. (in press) A century of evolution in *Spartina anglica. Advances in Ecological Research* **20**.

GROENENDIJK, A.M. (1986) Establishment of a *Spartina anglica* population on a tidal mudflat: a field experiment. *Journal of Environmental Management* **22**, 1–12.

HUBBARD, J.C.E. (1969) Light in relation to tidal immersion and the growth of *Spartina townsendii* (s.l.). *Journal of Ecology* **57**, 795–804.

HUTCHINSON, G.E. (1957) Concluding remarks. *Cold Spring Harbour Symposium in Quantitative Biology* **22**, 415–427.

KERSHAW, K.A. (1976) The vegetational zonation of the East Pen Island saltmarshes, Hudson Bay. *Canadian Journal of Botany* **54**, 5–13.

LETSCH, S.W. & FREY, R.W. (1980) Deposition and erosion in a Holocene saltmarsh, Sapelo Island, Georgia. *Journal of Sedimentary Petrology* **50**, 529–542.

LEVINS, R. (1968) *Evolution in Changing Environments: Some Theoretical Explorations*. Princeton University Press, New Jersey.

LONG, S.P. (1983) C_4 photosynthesis at low temperatures. *Plant, Cell and Environment* **6**, 345–363.

LONG, S.P. (1990) The primary productivity of *Puccinellia maritima* and *Spartina anglica*: a simple predictive model of response to climatic change. In J.J. Beukema, W.J. Wolff & J.J.W.N. Brouns (eds) *Expected Effects of Climatic Change on Marine Coastal Ecosystems*. Kluwer, Dordrecht, 33–39.

LONG, S.P., INCOLL, L.D. & WOOLHOUSE, H.W. (1975) C_4 photosynthesis in plants from cool temperate regions with particular reference to *Spartina townsendii*. Nature **257**, 622–624.

MACARTHUR, R.H. (1972) *Geographical Ecology: Patterns in the Distribution of Species*. Harper & Row, New York.

PETHICK, J.S. (1981) Long-term accretion rates on tidal saltmarshes. *Journal of Sedimentary Petrology* **51**, 571–577.

PETHICK, J.S., LEGGETT, D. & HUSAIN, L. (1990) Boundary layers under saltmarsh vegetation developed in tidal currents. In J.B. Thornes (ed.) *Vegetation and Erosion*. John Wiley & Sons Limited, London, 113–123.

PIANKA, E.R. (1974) Niche overlap and diffuse competition. *Proceedings of the National Acadamy of Sciences, USA* **71**, 2141–2145.

PIELOU, E.C. & ROUTLEDGE, R.D. (1976) Saltmarsh vegetation: latitudinal gradients in zonation patterns. *Oecologia* **24**, 311–321.

RANDERSON, P.F. (1979) A simulation model of saltmarsh development and plant ecology. In B. Knights & A.J. Phillips (eds) *Estuarine and Coastal Land Reclamation and Water Storage*. Saxon House, Farnborough, 48–67.

RANWELL, D.S. (1972) *Ecology of Saltmarshes and Sand Dunes*. Chapman and Hall, London.

RUSSELL, P.J., FLOWERS, T.J. & HUTCHINGS, M.J. (1985) Comparison of niche breadths and overlaps of halophytes on saltmarshes of differing diversity. *Vegetatio* **61**, 171–178.

SCHOLTEN, M. & ROZEMA, J. (1990) The competitive ability of *Spartina anglica* on Dutch saltmarshes. In A.J. Gray & P.E.M. Benham (eds) *Spartina anglica – A Research Review. Institute of Terrestrial Ecology Research Publication No. 2*. HMSO, London, 39–47.

SCOTT, R. CALLAGHAN, T.V. & LAWSON, G.J. (1990) *Spartina* as a biofuel. In A.J. Gray & P.E.M. Benham (eds) *Spartina anglica – A Research Review. Institute of Terrestrial Ecology Research Publication No. 2.* HMSO, London, 48–51.

SILANDER, J.A. & ANTONOVICS, J. (1982) Analysis of interspecific interactions in a coastal community – a perturbation approach. *Nature* **298**, 557–560.

VAN EERDT, M.M. (1985) The influence of vegetation on erosion and accretion in saltmarshes of the Oosterschelde, The Netherlands. *Vegetatio* **62**, 367–373.

WIT, C.T. DE (1960) On competition. *Verslagen van Landbouwkundige Onderzoekingen* **66**, 1–82.

5

The conservation of British saltmarshes

J.P. DOODY

Introduction

Saltmarshes are valued in a variety of ways depending on the perspective from which they are viewed. To the agriculturalist they can provide high quality farming land, when 'ripe for reclamation'; to those concerned with development they provide opportunities for creating additional land suitable for building. The ecologist may see their value in providing a relatively simple example of primary succession. To the bird watcher their attraction lies in the wintering birds that may roost or feed on them. Some of these birds in their turn provide the wildfowler with something to shoot. From a recreational or landscape point of view they are described as 'wilderness' or not considered at all.

To the conservationist they provide a very different picture. The saltmarsh itself is made up of plants adapted to the rigours of the salt-laden environment and frequent inundation by the tide which are not found in other habitats. As the saltmarsh develops, temporal and spatial variation in the vegetation increases along with the component animals. Where there are transitions to other habitats sometimes rich associations occur.

Saltmarshes also exist in a mosaic of other habitats, including tidal waters, intertidal flats, sand dunes and shingle bars. The complex patterns of creek development and pan formation, the cycles of erosion and deposition and the daily movement of the tides all suggest a richness and diversity which goes beyond the immediate confines of the saltmarsh itself. In estuaries these habitats combine to form essential locations for many species of birds of national and international significance.

Conserving this complex and often changing environment requires an understanding not only of the ecological principles and interactions which help to form the saltmarsh but also the anthropogenic impacts upon it. This paper attempts to describe the historical perspective from which saltmarshes are viewed by those exploiting them, to identify the more important nature conservation features, and to analyse some of the factors which are important to their continued survival.

Historical perspectives

The oldest form of exploitation by man is probably that associated with the development of saltmarsh for agriculture, with the earliest uses being grazing, samphire gathering or hay-making. However, in Britain at least, the building of an earth bank to exclude the tide has also been practised for centuries. Perhaps the best documented example is The Wash in Lincolnshire. The process begins with the building of a small 'cradge' bank to seaward of the location of the new outer wall. This provides a working area normally free from inundation by the tide. Material for the main sea wall is excavated from a 'borrow-dyke' which may be in front or behind the new bank depending on the width of saltmarsh to seaward. By this method extensive areas of land have been 'won' from the sea since Saxon times (Dalby, 1957). Enclosure of saltmarsh often takes place on a piecemeal basis. However, over long periods of time substantial areas can be enclosed, as has happened in many of our major estuaries (Table 1).

Historically, following enclosure the saltmarshes were used as coastal grazing marsh, which helped create important habitats in their own right. Traditionally managed permanent pastures, which may be periodically flooded in winter, can provide important nesting and winter feeding areas for birds. In addition, within the associated ditches, borrow-dykes and relict saltmarsh creeks there is often a rich flora and invertebrate fauna. Important areas include the Somerset and Gwent Levels in South West England and South Wales respectively and large areas in Essex and Kent. In recent years, however, many of these pastures have been converted to arable land. In the Thames Estuary, for example, nearly 70% of the grazing marsh has been lost in this way since the last war (Thornton and Kite, 1990). This process not only destroys the grassland, but also results in a decline in the interest of the ditches as a consequence of the more intensive management of the 'improved' fields and the resulting eutrophication.

This further loss completes the process of the conversion of tidal land to high intensity agricultural use. In more recent years this

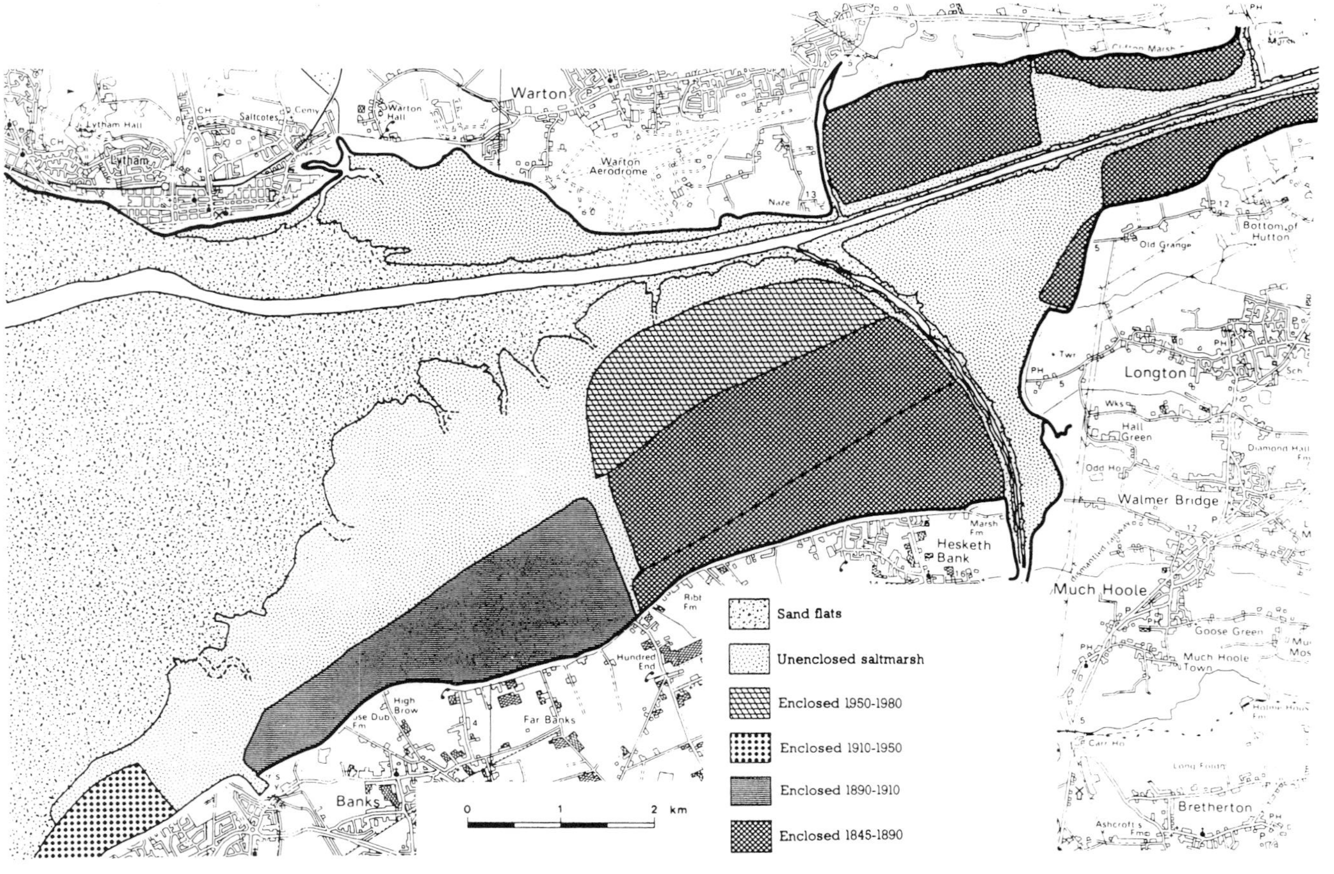

Fig. 5.1. The enclosure of saltmarsh for agricultural development in the Ribble Estuary, 1845–1980.

Table 5.1. *Major areas of historical 'reclamation' of saltmarsh for agriculture.*

Geographical area	Area affected (Dates in ha)	Reference
Essex and North Kent	4340 (mainly pre 18th Century)	(Macey, 1974)
'The Wash' (Lincs/Norfolk)	29,000 (since 17th Century) 3000 (in 20th Century)	(Dalby, 1957)
Morecambe Bay (Cumbria/Lancashire)	1300 (13th Century to 19th Century)	(Gray, 1972)
The Ribble (Lancashire)	1960 (last Century)	(Royal Commission on Coastal Erosion, 1911)
The Humber (Humberside)	4663 (17th to mid 19th Century)	(Stickney, 1908)
Other sites referred to in the third (and final Report of the Royal Commission on Coastal Erosion (1911): reclamation for agriculture and industry.		
Dee Estuary (Cheshire/ Clwyd)	3160 (by 1857)	
Severn	*c.* 8000 (by the Romans and subsequently)	
Mersey	492 (last Century)	
Firth of Forth	250 (since 1820)	
Firth of Tay	149 (in the 19th Century)	
Nigg Bay	80 (in the 19th Century)	

intermediate stage has been bypassed and the land put into arable production within just a few years. In the case of the Ribble Estuary (Fig. 5.1) the last enclosure took place as recently as 1980. Here an earlier loss of land designated as a Site of Special Scientific Interest was only prevented by the intervention of the Government, which provided money for the Nature Conservancy Council to buy it.

The infilling of tidal flats and saltmarshes (e.g. for waste disposal) is a more recent activity and, although on a smaller scale, has also had a considerable impact on some of our estuaries. Examples include Teesmouth, where over 2000 ha of intertidal land within the estuary had been developed by 1974 for port facilities, oil refineries and a power station (Evans and Pienkowski, 1984) and Southampton Water, where port and other related developments have claimed 1090 ha of the estuary (Tubbs, 1981).

The combined effect of these two forms of enclosure of tidal land has resulted in the loss of considerable areas of some of our major estuaries. In particular, in so far as the saltmarsh is concerned, this has resulted in a truncation of the zones with the destruction of the upper marsh and transitions to other habitats, notably fen and reedswamp. It is against this background that man's further use of saltmarsh must be judged, since it represents an already depleted resource. Although the remaining areas still display many important characteristics of a 'natural habitat', these can be highly modified at some sites. As we shall see below, the process of saltmarsh enclosure has important implications, not only for the saltmarshes themselves but also for nature conservation interests, which go far beyond the immediate impact of the area affected.

Nature conservation significance

There are some 44,000 ha of saltmarsh in Great Britain (Fig. 5.2). Estuarine saltmarshes are the most extensive. Large quantities of silt are normally present which, together with the relatively sheltered nature of the intertidal banks, provide ideal conditions for rapid sedimentation and saltmarsh growth. Some of the largest expanses of saltmarshes are to be found in sites like The Wash (4133 ha), the Solway Firth (4219 ha) and Morecambe Bay (3077 ha). On more exposed coasts, where wave energy is higher and sediment load lower, saltmarshes develop behind wide intertidal sand flats or in the shelter of offshore barrier islands (e.g. North Norfolk coast and Culbin, northeast Scotland). Together these make up some 95% of the resource and are distributed around the whole coastline, with major concentrations in southeast and northwest England (Burd, 1989) (Fig. 5.2).

Saltmarsh may also develop in areas where there is little or no sediment. In these circumstances, it becomes established in response

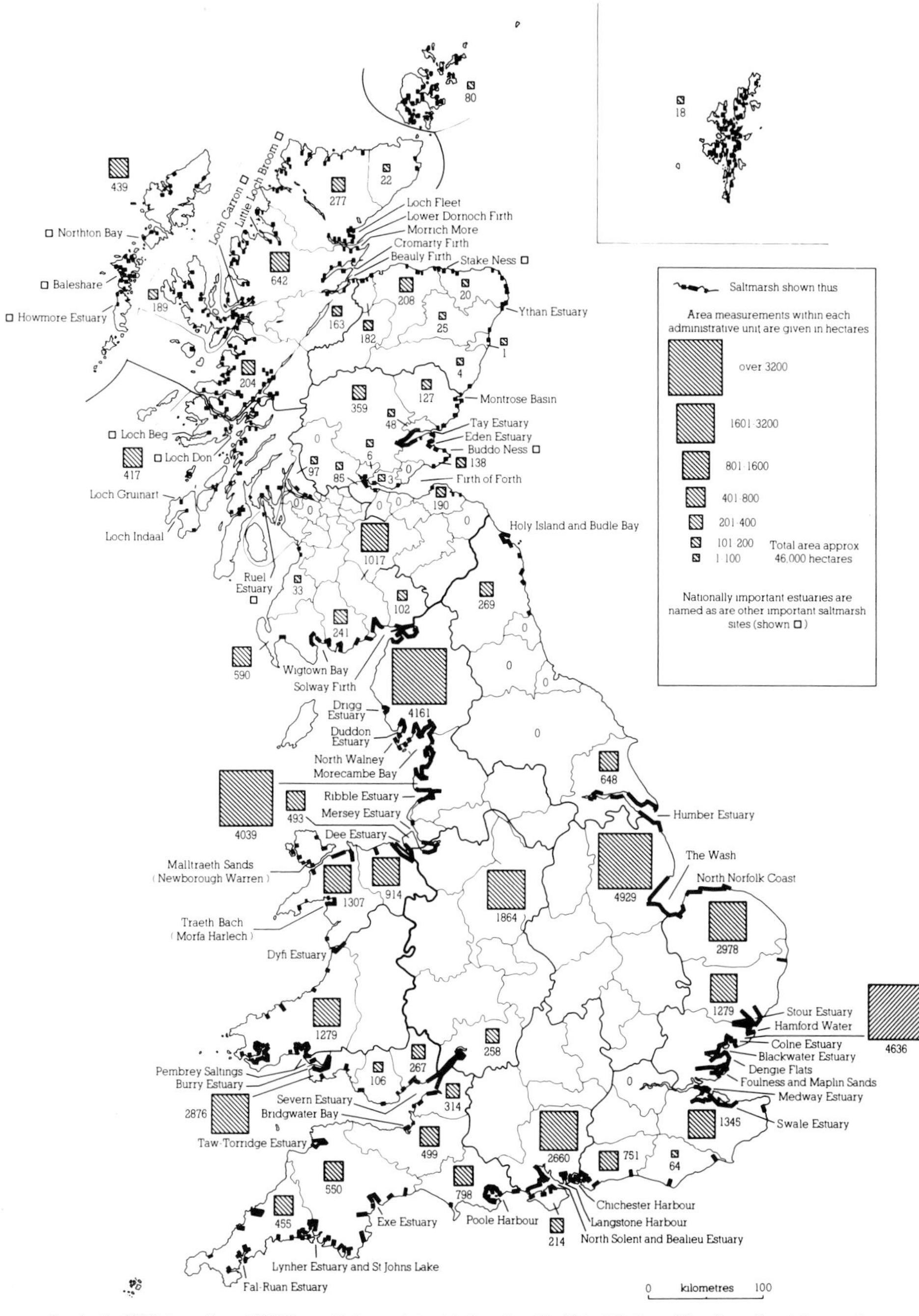

Fig. 5.2. Distribution of saltmarshes larger than 0.5 ha in Great Britain.

to regular tidal inundation, though in the absence of sediment accretion the communities tend to become fixed in a spatial sequence which changes little with time. The best examples of these are found in the western sea lochs or on rocky beach plains in Scotland.

Saltmarsh communities can also occur on cliff-tops (especially in Scotland where winter storms may drive salt spray well above high water), inland in areas with underground salt deposits, and behind sea walls where sea-water intrusion takes place in old creeks and drainage dykes. These comparatively minor developments were not covered in the survey referred to above and are only briefly considered in this paper.

Vegetation communities Saltmarsh vegetation tends to develop in roughly the same sequence. Lower levels are more frequently inundated with salt water and support pioneer communities of glasswort (*Salicornia europaea*), annual sea-blite (*Suaeda maritima*), sea-aster (*Aster tripolium*) and cord-grass (*Spartina anglica*) individually or in combinations. As the frequency of tidal immersion decreases, either as accretion takes place or as the physical character of the shoreline changes, additional species appear.

The communities which develop above the pioneer zone are usually characterised by the presence of sea poa (*Puccinellia maritima*) at the low-mid level, red fescue (*Festuca rubra*) at the mid to upper levels, with saltmarsh rush (*Juncus gerardii*) at the higher levels reached by spring tides. Each of these species forms a major component of the vegetation, and the naming of the vegetation types in the National Vegetation Classification (Rodwell, in press) reflects their importance. SM13, the *Puccinellia maritima* saltmarsh, and SM16, the *Festuca rubra* saltmarsh (which includes the communities with *Juncus gerardii*), are by far the largest and most complex groups in Great Britain.

In the absence of reclamation transitions to upper marsh with common reed (*Phragmites communis*) and sea club-rush (*Scirpus maritimus*) are found, particularly in the upper parts of estuaries. Where saltmarshes abut transitions to other natural and semi-natural habitats, vegetation develops which may support rich fen communities particularly where there are fresh water seepages into the marsh. These areas may be very rich in plants and invertebrates. In a very few locations trees may be found growing at the edge of the marsh. Table 5.2 gives a summary of the classification of the vegetation according to the National Vegetation Classification.

Table 5.2. *List of plant communities of British saltmarshes as defined by the National Vegetation Classification (Rodwell, in press).*

NVC Communities	
SM4	*Spartina maritima* saltmarsh
SM5	*Spartina alterniflora* saltmarsh
SM6	*Spartina anglica* saltmarsh
SM7	*Arthrocnemum perenne* stands
SM8	Annual *salicornia* saltmarsh
SM9	*Suaeda maritima* saltmarsh
SM10	Transitional low marsh vegetation with *Puccinellia maritima*, annual *Salicornia* species and *Suaeda maritima*
SM11	*Aster tripolium var. discoideus* saltmarsh
SM12	Rayed *Aster tripolium* as saltmarshes
SM13	*Puccinellia maritima* saltmarsh
SM14	*Halimione portulacoides* saltmarsh
SM15	*Juncus maritimus/Triglochin maritima* saltmarsh
SM16	*Festuca rubra* saltmarsh
SM17	*Artemisia maritima* saltmarsh
SM18	*Juncus maritimus* saltmarsh
SM19	*Blysmus rufus* saltmarsh
SM20	*Eleocharis uniglumis* saltmarsh
SM21	*Suaeda vera/Limonium binervosum* saltmarsh
SM22	*Halimione portulacoides/Frankenia laevis* saltmarsh
SM23	*Spergularia marina/Puccinellia distans* saltmarsh
SM24	*Elymus pycnanthus* saltmarsh
SM25	*Suaeda vera* saltmarsh
SM26	*Inula crithmoides* on saltmarshes
SM27	Ephemeral saltmarsh vegetation with *Sagina maritima*
SM28	*Elymus repens* saltmarsh

The many factors which help to modify the type of saltmarsh which develops on a particular site provide the basis for sometimes distinctive geographical relationships within the vegetation (Adam, 1978). A description of the three most significant areas is given below. Table 5.3 presents some indication of the more natural distributions of species forming important components of British saltmarshes. *Spartina anglica*, a main component of many saltmarshes in England and Wales, is not included in the Table. However, since its appearance on the south coast of England, following the crossing of the

Table 5.3. *Geographical distribution of important saltmarsh plants in Great Britain.*

a. Species generally of south and south-east England

Frankenia laevis	*Salicornia perennis*
Suaeda fruticosa	*S. pulsilla*
Limonium bellidifolium	*Spartina maritima*
L. binervosum (saltmarsh only)	*S. alterniflora*
Puccinellia fasciculata	*Inula crithmoides* (saltmarsh only)
P. rupestris	

b. Species found in England and Wales but absent north of the Solway

Halimione portulacoides
+*Artemisia maritima*
+*Limonium vulgare*
L. humile

+ Extend further north in scattered localities on the east coast.

c. Species with a generally southern distribution but absent from the extreme north of Scotland

Juncus maritimus (a generally western species at its northern limit)

**Scirpus maritimus*	
Schoenoplectus tabernaemontani	
**Oenanthe lachenallii*	(a western species predominantly, particularly at its northern limit)
Atriplex littoralis	(a species with a generally southern and eastern distribution)

d. Species with a predominantly northern distribution

**Blysmus rufus*
Carex recta
**Juncus balticus*
**Carex maritimus*

* Species not restricted to saltmarshes.

native *Spartina maritima* with an introduced *Spartina alterniflora*, it has had a major impact on saltmarsh development (Doody, 1985; Gray and Benham, 1990).

The geographical area of south and southeast England supports some of the floristically richest saltmarshes in Great Britain. These are found on high sandy beach plains in association with sand dunes. The best examples occur on Scolt Head Island, Norfolk. They are ungrazed and support a number of rare species, including sea-heath (*Frankenia laevis*) and shrubby sea-blite (*Suaeda fruticosa*), with a generally Mediterranean distribution. Most of these saltmarshes include sections which are also free from enclosure so that the full floristic representation can be seen from low-level marsh to upper marsh and transitions to non-tidal vegetation. Locally on the mainland shore the transition to *Phragmites* is particularly well developed.

Elsewhere in the southeast other marshes are ungrazed and, though subject to enclosure and conversion to agricultural land, particularly in Essex and Kent, support good examples of *Halimione portulacoides* dominated communities. The visually distinct community with *Limonium vulgare* is also well represented.

The presence of extensive pioneer marsh – notably sea-blite (*Suaeda maritima*), glasswort (*Salicornia europea*) and sea-aster (*Aster tripolium*) – is also a feature in this region. Saltmarsh enclosure has tended to favour the development of pioneer marsh as accretion is stimulated in front of any new sea bank. This is particularly well exemplified in The Wash, where some of the most recent reclamations have taken place.

Townsend's cord-grass (*Spartina anglica*) is also a notable feature of many of the south coast marshes where it forms a major component of the vegetation at many sites. In some areas it is thought to have replaced the native *S. maritima*. Its origin, rapid expansion and 'natural' demise from sites on the south coast has been noted by several authors, e.g. Poole Harbour (Gray and Pearson, 1985) and Langstone Harbour (Haynes and Coulson, 1982).

One important physical feature of these marshes, particularly in Essex and Kent, is the extent to which they are eroding at their seaward edge. This appears, partly at least, to be a response to the relative rise in sea level, which has been estimated at approximately 3 mm annually.

The saltmarshes of the southwest coasts are very restricted although they do occur in the small estuaries of Devon and Cornwall.

The saltmarshes of western England (northward from Bridgewater Bay) and Wales tend to be grazed, by comparison with south and southwest England. Here the extensive grassy lawns, so familiar to ornithologists, provide important feeding areas for grazing ducks and geese. The heavy stock grazing pressures normally encountered tend to restrict plant species diversity, particularly on marshes that are also used for turf cutting, and eliminate any structural diversity. At a few sites, such as Rockliffe marsh on the Solway, lower grazing levels allow more structure to develop, providing important habitat for breeding birds. Morecambe Bay also supports transitions to grassland which has little structural diversity but retains a reasonable species complement. On some grazed marshes characteristic *Juncus maritimus* communities survive which, in at least one site (Burry Inlet, South Wales), include good stands of the marsh mallow (*Althaea officinalis*).

Elsewhere, though rather more restricted in their distribution, traditionally ungrazed marshes occur, particularly in South Wales. Here transitions to sand dunes are also present and the species complement has important affinities with north Norfolk, notably with the presence of rock sea-lavender (*Limonium binervosum*) and golden samphire (*Inula crithmoides*) in the upper marsh. Although there has been reclamation within some of the larger estuaries, upper saltmarsh is well represented, particularly that dominated by *Juncus maritimus*. The incidence of erosion is less evident here, possibly because relative sea level is either comparatively stable or is rising less rapidly than in southeast England.

Spartina anglica is an important component of the vegetation at a number of sites and there is evidence that it is currently going through a rapid phase of expansion at others. In the Dovey, the Dee and the Ribble estuaries, there are already some 938 ha of the species (Burd, 1989) and expansion is well documented, for example, in Morecambe Bay (Whiteside, 1987).

In floristic terms the main saltmarsh communities of Scotland are impoverished by comparison with their southern counterparts. In particular, a number of the more important components of the marshes of England and Wales – notably *Limonium* spp., *Halimione portulacoides* and *Artemisia maritima* – are virtually absent north of the Solway Estuary (Table 5.3). However, despite this, and the fact that Scottish marshes tend to be scattered and small by comparison with those in England and Wales, they are important in other ways. In particular, the incidence of reclamation is much less evident and this combined with the high rainfall has helped to ensure complete transitions to non-tidal vegetation. One site in the southwest of the head of the Ruel Estuary includes upper saltmarsh with scattered alder and represents one of the very few natural transitions to 'woodland'. Typically, at many sites around the coast, these upper marsh zones include *Iris pseudacorus* as a major component.

Table 5.4 *Saltmarsh fauna in Britain (provisional list compiled in 1979).*

	Total resident species	Total exclusive to saltmarsh
Lepidoptera (moths)	50	24
Coleoptera (beetles)	106	47
Diptera (flies)	58	43
Hymenoptera (bee and sawfly)	2	2
Hemiptera (bugs)	66	26
Orthoptera (grasshoppers etc)	3	—
Aranaea (spiders)	8	6
	293	148

Fauna Current information on the invertebrates of saltmarshes is relatively limited (Stubbs, 1982). Hence, only general comments can be made on the relationship between the saltmarsh plant species and their importance to the invertebrate fauna.

Table 5.4 indicates the total number of species within a few of the major groups of invertebrates found regularly on saltmarshes and the number which are believed to be confined to them. Many more species can occur on saltmarsh as strays from other habitats or as flower visitors from terrestrial breeding sites.

Diptera and Coleoptera are the two predominant orders but moths, bugs and other insects also occur, together with spiders and mites. It is of interest that some species are specific to saltmarsh, while others are surprisingly adaptable and include saltmarsh among a range of terrestrial habitats. The latter trait is well exemplified by the moth *Aphelia vibernana* which on moorland feeds on such plants as bilberry, whereas the bug Aphrodes bicinctus has a distinct sub-species confined to saltmarsh.

About one third of the fauna is phytophagous (plant eating). Since the best plants are *Aster tripolium*, *Artemisia maritima* and *Limonium* spp. It follows that there is likely to be a good invertebrate fauna where these plants are present. The importance of an individual saltmarsh depends not only on structural diversity but also in the presence of the above species. Since grazing adversely affects both of these, the overall value will be diminished as grazing pressure increases. Transitional communities also provide important habitat and amongst these *Phragmites* stands associated with freshwater seepages into saltmarsh are particularly valuable. Good examples of these may be found throughout Britain, particularly where there has been little or no saltmarsh enclosure.

Breeding birds on saltmarshes are restricted to a few species. In lowland areas, where species which formerly bred on unimproved meadow land have been displaced by intensive agricultural use, they provide an important habitat. Amongst these are redshank, which breed on 82% of 77 representative sites surveyed, oystercatchers (73% of sites) and lapwing (33% of sites: Allport *et al.*, 1986). Nesting densities as high as 80 nests per km^2 have been estimated for redshank in saltmarshes in The Wash (Green *et al.*, 1984). Passerines are also restricted in species representation on saltmarshes. However, skylarks, meadow pipits and reed bunting are commonly found as breeding species (Fuller, 1982).

Wintering birds include large numbers of herbivorous ducks and geese which find the shorter and more palatable grasses such as *Puccinellia maritima* and *Festuca rubra* suitable for grazing. Other species, such as the twite on The Wash (Davies, 1987), can attain significant numbers and the roosting area provided by some of the larger saltmarshes for species such as the short-eared owl may be significant.

Saltmarsh productivity Saltmarshes exist in a complex environment and in the estuarine situation represent an important component in the functioning of the system. Figure 5.3 illustrates a highly simplified representation of the interaction between the various components. Of particular interest to the ecologist is the extent of primary productivity which can be attributed to the saltmarsh. The traditional view is that this is high and a major contribution to estuarine productivity. However, the situation is far more complex than this, and the extent of below-ground productivity and the effect of grazing are among the factors which can affect the relative contribution from the saltmarsh vegetation (Long and Mason, 1983). In addition, the interchange with the tidal waters, including the contribution from other primary producers, and the amount of detritus from external sources, can make up a significant proportion of the overall productivity.

All of the above factors are important when considering the saltmarsh habitat and its value from a nature conservation point of view. The assessment of the importance of an individual marsh is

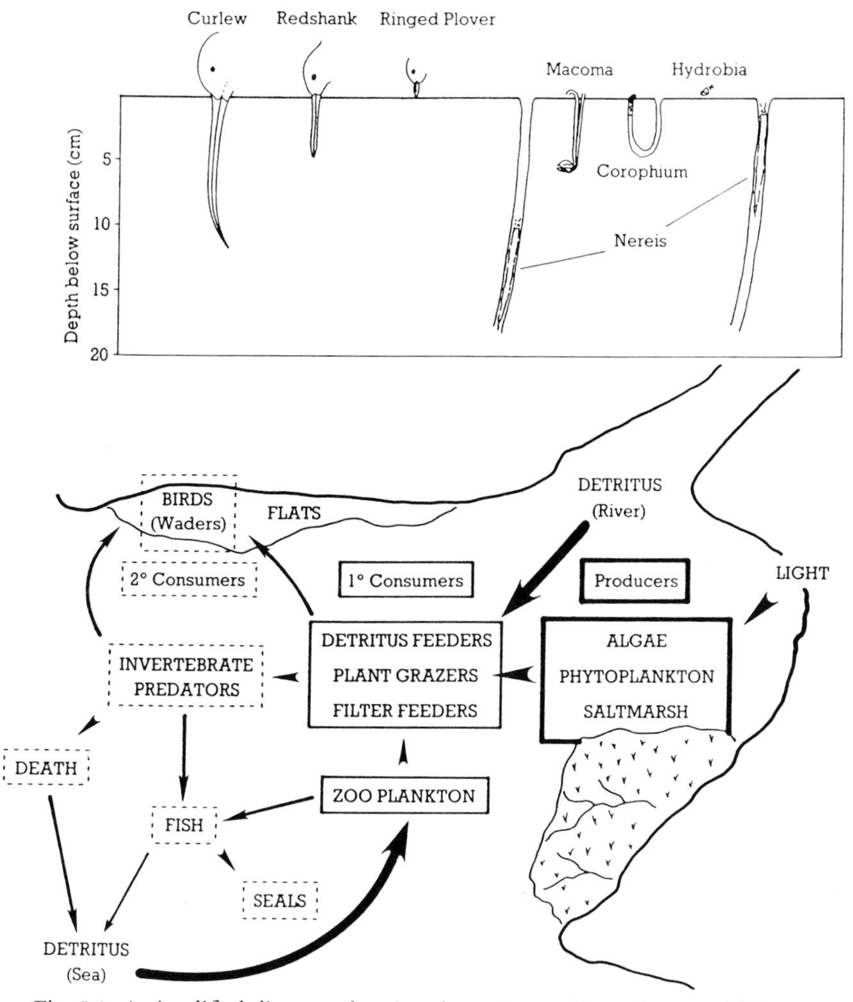

Fig. 5.3. A simplified diagram showing the main trophic pathways within an estuary.

based on detailed guidelines. These are discussed in more detail in the next section which includes a summary of the main methods of protection.

Site protection

The selection of saltmarshes for protection as Sites of Special Scientific Interest (SSSI), under the Wildlife and Countryside Act

Table 5.5 *A summary of the habitat selection units for saltmarsh Sites of Special Scientific Interest. See Table 5.2 for list of saltmarsh vegetation types, SM 1–27.*

Zostera/Ruppia low-marsh – SM1–3 *Zostera* spp., *Ruppia maritima* and communities with *Eleocharis parvula*.

Pioneer marsh – includes SM4–6 *Spartina*, SM7–9 *Salicornia* (including *Arthrocnemum*)/*Suaeda* stands and SM11 and SM12 *Aster* stands.

Low-mid marsh – includes SM10 transitional low marsh, the sub-community of SM13 with *Puccinellia* dominant and all three SM14 *Halimione* communities.

Mid-upper marsh – includes sub-communities of SM13 which have *Limonium, Armeria, Plantago maritima* and *Glaux* prominent, SM16 *Festuca* communities, SM17 *Artemisia*, SM19 and SM20 communities of wet depressions and SM15 and SM18 with *Juncus maritimus*.

Drift-line – SM24 and SM28 *Elymus* communities and SM25 *Suaeda vera*.

Swamps – S4 and S19–21 (covered in the NVC classification of swamp vegetation: see also C7); may form important upper marsh or upper estuary transitions.

Transitions – include SM21 and SM22 dune transitions, MG11–13 freshwater transitions (covered in the NVC classification of mesotrophic grassland and M28 mire transitions (Rodwell, in press).

1981, is based on guidelines which cover many of the aspects mentioned above. However, the geographical representation of saltmarsh vegetation described above, which uses the National Vegetation Classification as the basis for identifying the sequence of communities within individual marshes, is a major part of the selection procedure. Table 5.5, taken from the Nature Conservancy Council's Guidelines for the Selection of Biological SSSIs, gives the main breakdown for the units chosen.

The best sites are those which include a full sequence of vegetation types, provide good examples of each, and occur in combinations which are representative of their geographical position. Additional interests – wintering birds, breeding birds, rare plants, invertebrates, and amphibians and reptiles – are also important and included in other sections of the guidelines.

Until 1981 the primary legislation giving protection to important sites for nature conservation (identified as Sites of Special Scientific Interest) mainly covered activities requiring planning permission. This designation helped to protect sites, including saltmarshes, from destruction by roads, barrages and industrial development. However, many other developments including embankment and reclamation for agriculture, changes in grazing regime and drainage, which were classed as agricultural activities, fell outside the provisions of the 1949 National Parks and Access to the Countryside Act under which Sites of Special Scientific Interest were formerly designated.

The 1981 Wildlife and Countryside Act (Section 28) requires the NCC not only to identify and notify the Local Authority of Sites of Special Scientific Interest as had been done under the 1949 Act, but also owners and occupiers of the land and the Secretary of State of its intention to do so. In notifying the site the Nature Conservancy Council also has a duty to identify potentially damaging operations. These are operations which generally fall outside planning legislation but potentially, at least, could cause loss of scientific interest of the site. Those which apply most often to saltmarshes and are likely to have greatest impact include the following:

a. Drainage of the marsh surface and changes in the form of tidal creeks. These may affect the flow of water over the saltmarsh and are important elements in the tidal dispersal and deposition of sediment. Changes in the drainage pattern can have important consequences for erosional and depositional phases in saltmarsh development.
b. Digging of scrapes and tidal ponds. Additional feeding or nesting habitat for birds may be provided by these developments. However, there will be a loss of saltmarsh vegetation and its component fauna. The precise impact of these activities will depend on their size, permanence and the original interest of the marsh.
c. 'Reclamation' of land from sea, estuary or marsh. As noted above, this is the single most damaging activity affecting saltmarshes, truncating the sequence of development and impoverishing their flora and fauna.
d. Improvement of sea defences. When this involves the excavation of saltmarsh in front of a sea wall to reinforce and heighten it, this may destroy quite large areas of upper marsh.
e. Changes in grazing management (see below).

In attempting to protect saltmarshes from any of the above activites, or indeed any other operations which would adversely affect their wildlife interest, the Nature Conservancy Council can enter into a management agreement on any SSSI. This may involve the payment of compensation for not carrying out an activity (e.g. embankment and reclamation), or the change of a management regime which might include compensation for a reduction in the number of grazing stock kept on a marsh.

Burd (1989) estimated that more than 80% of saltmarshes lie within sites protected in this way. Although this represents a high proportion of the total habitat, the largest areas lie within estuarine sites identified primarily for their ornithological interest. This is not to diminish the significance of the saltmarsh but is more a reflection of the bias in site selection procedures, particularly at international level, which are based on total numbers of individual waterfowl using a site in winter. These criteria have gained widespread recognition and are embodied in international legislation associated with the identification of sites worthy of protection under the RAMSAR convention and the EC Directive on the Conservation of Wild Birds.

One of the most effective ways of protecting sites for nature conservation is to own and/or manage them as nature reserves. In this way, theoretically at least, all conservation needs are catered for since this is recognised as the primary objective of land management. However, this implies an adequate control over management operations and, more importantly, an ability to define the conservation aims. It will emerge below that there are a number of different options involving assessment of the recent history of man's use or abuse of the site.

National Nature Reserves owned or managed by the Nature Conservancy Council and its successor bodies, which include saltmarshes as a main component of the designated area, are found throughout Great Britain. Figure 5.2 shows the location of nationally important sites in relation to the distribution of saltmarsh. All of the sites may be considered to be of national importance as saltmarsh habitat either because they are significant in their own right (e.g. large, rich in plants and animals and/or representative of important geographical variation) or because they form an integral part of a large site designated for other interests, notably birds (Ratcliffe, 1977). As such they are eligible for designation as NNR.

To date there are 16 sites which have significant areas of saltmarsh included within a designated National Nature Reserve, the most important being Scolt Head and Holme in north Norfolk, Hamford Water, Colne and Blackwater estuaries in Essex, the Ribble estuary in northwest England, and Caelaverock on the Solway Firth, southwest Scotland.

The Royal Society for the Protection of Birds (RSPB) is the primary voluntary conservation organisation whose land holding includes saltmarshes. To date there are in excess of 3000 ha of saltmarsh within 22 RSPB reserves. Most are at estuarine sites where the primary interest is the protection of wintering waterfowl.

The Voluntary Trust movement including the National Trust and the local Naturalist's Trusts also own or manage sites which include saltmarsh though no information is readily available on the areas involved.

Uses and abuses

There are a number of different uses to which saltmarshes are put whether they are protected as a Site of Special Scientific Interest or within a National Nature Reserve or other sites. These have varying impacts on their nature conservation interest depending on the scale and intensity of that use. Each of the main activities is described below in order to give an indication of their range and the way in which the saltmarsh is viewed by those exploiting it. A more detailed discussion of the most important issues affecting nature conservation is given later.

The use of saltmarsh for the grazing of domestic stock and turf cutting are the two main commercial uses, though the gathering of samphire and hay-making are practised in a very limited way (Beeftink, 1977). Currently grazing is the single most important management operation affecting a large number of saltmarshes in Great Britain (see below).

Grazing levels which typically produce the close-cropped lawns of the west coast are given as 4.75 sheep per ha, rising to 6.45 sheep per ha (Gray, 1972). Grazing is all year round except for the 80–100 days each year when high tides prevent all of the marsh being grazed. The short swards which are created by grazing not only provide important areas for stock feeding and grazing ducks and geese but may also be important in a few locations for the production of lawn turf (Gray, 1977).

Turf cutting areas are prepared by reseeding and the use of

fertiliser, such that the swards are composed of a very restricted number of species. This activity also has the effect of favouring *Puccinellia maritima*. Since good quality turf requires a high proportion of *Festuca rubra* there is a tendency for turfed areas to become commercially less valuable with time. Paradoxically, the reversion of the sward to one with a greater proportion of *Puccinellia maritima* results in it becoming more palatable to grazing geese.

The short close-cropped saltmarsh turf can provide an important food resource for some wildfowl species, and this is important in considering the most appropriate form of conservation management for individual sites. Traditionally, duck (notably wigeon) and geese have been shot and wildfowling continues in some areas. These include National Nature Reserves where shooting schemes are operated which form part of the mangement of the regime.

Enclosure of saltmarsh is accompanied by the erection of a sea bank. In a flat alluvial landscape this represents a long-term commitment to the prevention of innundation by the sea. That these defences can be very vulnerable was amply demonstrated in 1953 when some 158,000 acres of East Anglia were flooded by sea-water on the night of 31 January (Barnes and King, 1953). The different approaches adopted in strengthening these defences reflects, to some extent, the value attached to the saltmarsh in forming part of the coastal defences.

On The Wash the need to raise the sea banks, in order to compensate for the present relative rise in sea level on the east coast of England, is accommodated by the excavation of saltmarsh in front of the sea bank, effectively continuing the process of saltmarsh loss referred to above. The material which is deposited on top of the existing bank normally includes vegetation from the upper more mature parts of the marsh and significant areas of some of the most diverse saltmarsh communities have been lost in recent years (Hill, 1988).

On the Essex coast, by contrast, saltmarsh is seen, potentially at least, to be an important element in sea defence by dissipating incident wave energy. Because of this, experiments are currently in progress to rehabilitate eroding saltmarsh at several locations, notably on the Dengie Peninsula, by the National Rivers Authority Anglia Region (see Holder and Burd, 1990). Randerson (1984) discusses the value of saltmarshes in relation to coast protection from a biologist's point of view and their value on the Gwent Levels is reported by Green (1984). It is clear from these reports that there is a more enlightened approach developing in relation to the value of saltmarshes as an important element in what is often perceived as man's fight with the sea. However, there is still some way to go before an appreciation of their role in this context is fully realised.

Saltmarshes are not normally associated with recreational activities. The dissected terrain, particularly in the larger sites, with the numerous creeks and tall vegetation, makes walking difficult. Locally, narrow fringing marshes can provide access to boat moorings. Bird watching is perhaps the only significant use though even here this is mainly from sea walls or hides overlooking the marsh.

In some areas, notably northwest England, the development of *Spartina anglica* saltmarsh was seen as a threat to recreational interests as the amenity beaches at Southport were invaded. Much money and time has been spent in controlling the marsh since initial experiments were undertaken (Truscott, 1985).

The landscape value of saltmarshes has been recognised by artists and the 'wilderness' quality is one of their more sought-after attributes. However, despite this there are few of the major areas of estuarine saltmarsh included either as designated Areas of Outstanding Natural Beauty or Heritage Coast by the Countryside Commission. The most notable exception is the north Norfolk Coast (Pye, Chapter 8, this volume).

The scientific use of saltmarshes for the study of natural ecological processes has long been established and the work of V.J. Chapman (1938–41) is worthy of special note. Their apparently simple structure and successional characteristics have been extensively used for the teaching of basic ecological principles. These studies and later text books all stress the importance of adaptation to environmental conditions associated largely with salt tolerance and submergence in which few species survive (see Ranwell, 1972; Long and Mason, 1983). However, the description of accretional processes leading to higher and more diverse communities of both plants and animals is only part of the story. Natural variation in the physiography of the marsh, aspects of consolidation, erosion and accretion, together with anthropogenic mangement, all influence the way in which the nature conservation interest develops. Thus when looking at the conservation requirements the apparently simple picture presented by the ecologist, whilst helpful, must be treated with caution, particularly when considering management issues.

Management

Saltmarshes, whether protected as a Site of Special Scientific Interest, a statutory National Nature Reserve, owned by a voluntary conservation body or not protected at all, require management if they are to survive the uses and abuses outlined above. Each of the activities described is seen from a relatively narrow perspective. The farmer is concerned mainly with stocking levels, but the wildfowler with access and shooting quarry species, which the ornithologist may wish to watch. The nature conservation manager will have to look not only at the inherent value of the marsh, but also at the way it has been managed in the past. The reconciliation of the competing demands, in order to protect or enhance the existing features of the saltmarsh to secure its long-term conservation, is the main aim. The options for grazing management are discussed below, in order to illustrate the complexities of the considerations involved.

The effect of grazing is complex and has a long history on British saltmarshes. Today, whilst there are many sites which are ungrazed, a large number of these show evidence of this former use. In order to identify the most appropriate form of grazing management it is crucial to have some knowledge of the recent grazing management history of an individual site (Fig. 5.4).

Lightly grazed or ungrazed saltmarshes are those where historically intermittent very low-level grazing by domestic stock has taken place or where natural herbivores (taken here to mean birds, rabbits and hares) are the only grazing animals and accordingly tend to support the most floristically important vegetation types. In addition, the presence of a good structural diversity together with the grazing sensitive species *Halimione portulacoides, Limonium vulgare* and *Artemisia maritima* as significant components of the sward, provides important habitat for invertebrates and breeding birds.

Saltmarshes which have a long history of heavy grazing by domestic stock tend to lack the features described above. Typically these sites support grazing by up to 6.5 sheep per ha (year round) or 2 cows per ha (summer). Under these grazing regimes there tends to be an impoverishment of the flora. Characteristically, the grasses *Festuca rubra* and *Puccinellia maritima* are favoured, and the grazing sensitive species mentioned above are reduced or eliminated from the sward. Occasionally, at some sites where transitions to grazing marsh above normal tidal limits occurs, a species-rich close cropped grassland may be present, as in the Solway Firth and Morecambe Bay.

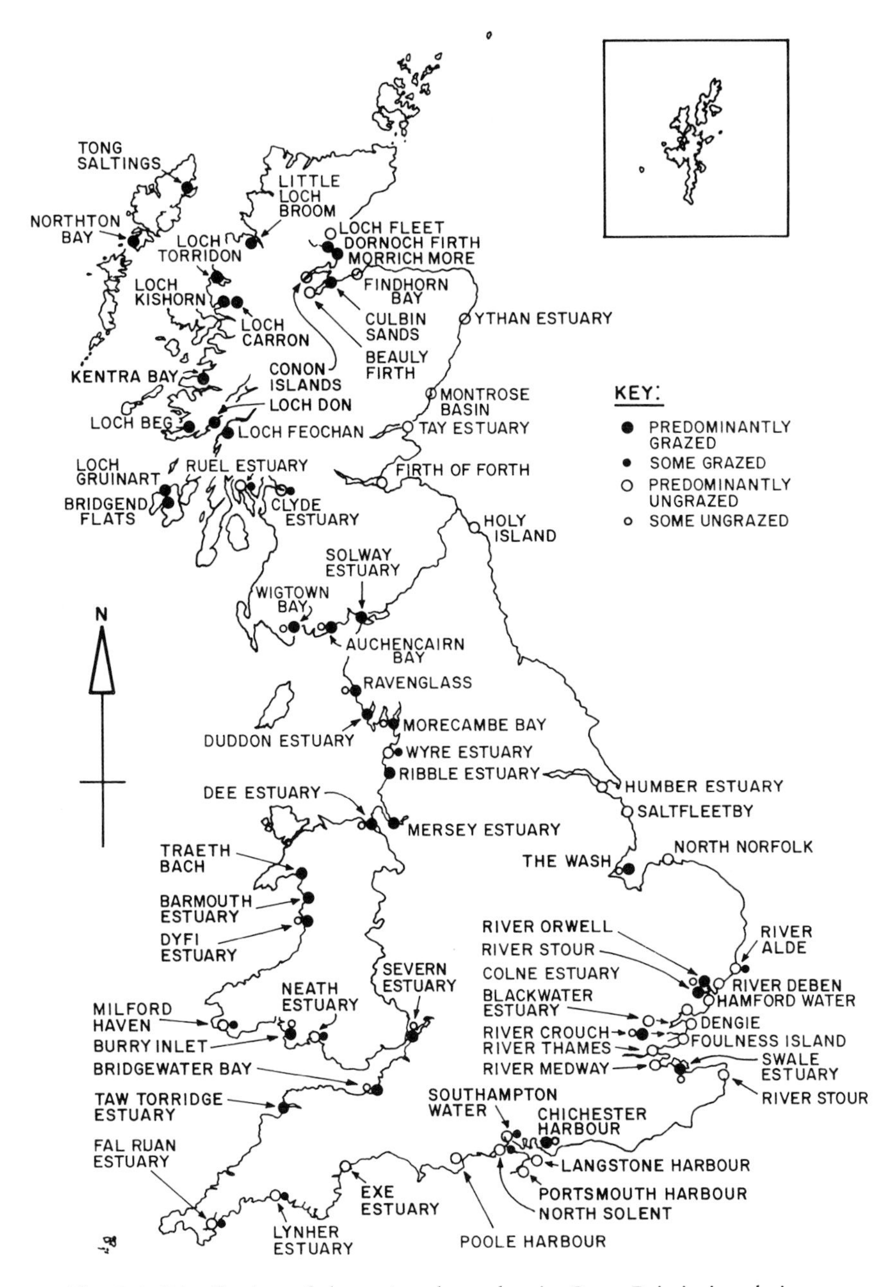

Fig. 5.4. Distribution of the main saltmarshes in Great Britain in relation to their grazing management.

The higher grazing levels result in the loss of structural diversity, which in its turn reduces the variety of invertebrates and breeding birds. Invertebrate interest is also further restricted by the absence of the main grazing sensitive plants which themselves provide food for a wide variety of invertebrates. The short-grazed swards, however, also support important populations of winter-grazing ducks and geese, often of national or international significance.

Moderately grazed saltmarsh, typified by grazing levels of approximately 0.33 cattle or ponies or 2 sheep per ha per year round, are thought to be most appropriate for nature conservation (Beeftink, 1977). On sites with a history of grazing at this level maximum structural diversity seems to be attained whilst at the same time retaining floral interest.

It can be seen that different levels of grazing can produce very different saltmarshes, in terms of both their structural diversity and species composition. This has a profound effect on the type of associated animal interest which develops. Lightly or ungrazed saltmarshes provide some of the most important natural examples of the saltmarsh habitats, particularly when found in association with sand dunes. As a general rule, continuation of the existing management regimes is the requirement.

At the other end of the spectrum, heavy grazing by domestic stock produces a short sward which usually favours grazing duck and geese at the expense of floristic and structural diversity. Whilst on most sites the ornithological interest may justify the continuation of the heavy stocking rates, at others consideration may be given to the value of increasing structural diversity, and this may lead to a decision to reduce grazing pressure. Whatever the positive effects of this action on breeding birds or invertebrates, there will be negative effects on wintering herbivores. In addition, it is clear from evidence of formerly grazed saltmarshes that a sudden reduction or elimination of grazing can have a profound effect on the vegetation. Rapid growth of the main floristic components takes place and a dense matted vegetation can develop in less than ten years. *Apropyron pungens* is known to be expanding in The Wash (Hill, 1987) and the Dee at the expense of more open saltmarsh vegetation as a result of reduced grazing pressure. *Festuca rubra* and *Puccinellia maritima* may react in the same way. This type of development is usually detrimental to the existing interest (by further reducing floristic diversity) and fails to provide alternative significant wildlife habitat.

Any decision to change grazing levels should therefore be made cautiously, with a gradual alteration in stocking density, and must be carefully monitored. Grazing levels, traditionally lower than those for moderately grazed saltmarsh, will tend to favour botanical and structural diversity whilst the opposite is generally true at higher levels.

Since the appearance of *Spartina anglica* on the south coast of England during the late 1800s some 12,044 ha of tidal flats have been covered with a monospecific sward (Hubbard and Stebbings, 1967). At many sites deliberate introduction by man has facilitated the expansion of the species and helped the reclamation process (Goodman *et al.*, 1959). Today there are several examples of rapid *Spartina* growth, not associated specifically with reclamation, which are giving cause for concern because of the loss of intertidal winter feeding ground for waterfowl. These include, notably, Lindisfarne NNR, Northumberland (Corkhill, 1980), Dyfi estuary NNR (Davis and Moss, 1985), and the Dee estuary (White, 1982). Other sites where rapid expansion is still taking place include the Mersey estuary, Ribble estuary and Morecambe Bay (Whiteside, 1987).

Spartina also impoverishes the saltmarsh from a botanical point of view. This is not usually by invasion into the existing marsh, but by taking on the mantle of the natural pioneers. The rapid growth and vigour of the *Spartina* sward tends to prevent the normal increase in species diversity as the marsh develops (Doody, 1990). Succession does take place on ungrazed sites to *Phragmites* (Ranwell, 1964) and to Puccinellia on grazed sites (Hill, 1987). In both cases one monospecific sward is replaced by another and there is usually little value in terms of plant species diversity.

Spartina may be important on eroding shores, providing protection for more mature marsh which would otherwise be lost. Consideration of this role poses difficult questions when reconciling the threat to nature conservation interests which its rapid expansion poses.

Discussion

The historical perspective shows us that saltmarshes have been exploited for a variety of purposes, most notably land claim, which causes its destruction. Enclosure of a saltmarsh (including tipping) and protection from tidal inundation results in the loss of the saltmarsh flora and its replacement by concrete or agricultural land.

This destruction represents a loss not only of the habitat and its associated plants and animals but also part of the functioning estuarine system. Conservationists have argued long and hard that these losses, however small, are cumulative (it is unusual for this type of reclamation to be a one-off exercise) and will ultimately threaten the well being of the whole site. The primary nature-conservation requirement is to prevent these destructive activities. This is a clear aim when considering the saltmarsh habitat if the full expression of zonation from pioneer to upper and transitional marsh is to be retained. Reclamation almost always destroys the upper, more mature marsh, effectively replacing a diverse, usually gradual transition, with a narrow, abrupt, artificial barrier.

The way in which these changes affect whole sites can be seen by looking at the examples of the Dee, The Wash, and the Ribble (Fig. 5.1). Enclosure of the upper Dee estuary began around 1730 and continued up to 1986 (Fig. 5.5a). Much of the land gained at the expense of the intertidal area has been developed for industry, housing and roads, though limited areas have also been used for agriculture including grazing marsh, the total amounting to approximately 4600 ha by 1877.

This gradual reduction of the size of the estuary has been followed by an increase in the extent of saltmarsh (Fig. 5.5b). This is thought to be partly due to the reduction in tidal volume, which in turn has reduced scour and helped accelerate the natural sedimentation tendency within the estuary. This process was greatly accelerated by the introduction of *Spartina anglica* around 1930 and since then some 1500 ha of intertidal mud has been naturally 'reclaimed'. Marker (1967) estimated that at Parkgate between 1947 and 1963 vertical accretion took place at a rate of approximately 25 mm per year.

These changes have also been partly responsible for the gradual relocation of the ports seaward. Up to about 1300 Chester, some 30 km from the mouth of the estuary, was a major port. As land claim and siltation increased, new ports were established as far down the estuary as Parkgate (Fig. 5.5a). Even this port did not last long and was abandoned in the 1820s. Subsequent growth of *Spartina* marsh sealed its fate and today one can look from the quayside across a vast expanse of saltmarsh which is only covered by the highest tides.

At the seaward edge of the marsh important changes also take place as a consequence of enclosure which may have implications for other interests within the estuary. Evidence suggests that there is a natural limit to the expansion of saltmarsh onto tidal flats (Kestner, 1962).

However, as each reclamation takes place the rate of accretion increases for a while in front of the new sea wall and pioneer marsh rapidly develops to seaward. Where the flats slope steeply towards low water mark there is little opportunity for accretion of new flats and as a result the intertidal area is diminished.

The combined effects of enclosure of the upper margins of the marsh and the sometimes rapid loss of tidal flats have important implications for the wintering populations of wildfowl and waders which rely for their food on invertebrate prey or *Zostera* and other plant material as the intertidal zone is squeezed. It is not clear when a diminishing intertidal zone may threaten the survival of the often internationally important bird populations. However, we know that birds use a succession of European estuaries during the winter period and that this may magnify the potential impact of any one reclamation. In the case of the Tees estuary, the loss of intertidal flats equated with a loss in the availability of invertebrate prey and a reduction in the number of birds using the site (Evans and Pienkowski, 1984). Although the precise significance of this to individual birds or populations of birds is difficult to predict, each being affected differently by the prevailing climate, and availability of preferred prey in other localities, the threat from these cumulative losses has been sufficient to sustain opposition to developments in The Wash, Cardiff Bay, Orwell and elsewhere in recent years.

It is against this background of habitat destruction that the nature conservationist's protectionist philosophy has therefore developed. However, the complex nature of the saltmarsh and its associated plants and animals, together with the significance in the functioning of the wider estuarine environment, require an appreciation of this complexity which goes far beyond the limited individual view of those who seek either to protect or to exploit it. It is the conservationist's role to try to reconcile competing demands, in order to protect the best examples of saltmarsh and try to ensure their proper integration within the complex of other habitats and uses in which they exist. This is no small task and the fact that saltmarsh loss is still one of the main impacts on our major estuaries suggests that a more enlightened approach to saltmarsh conservation is required.

The creation of new land for food production, or the building of houses, industry or roads was seen as a major gain for mankind. Certainly, without some of the major reclamation schemes, we would not have the extensive highly productive land around many of our major estuaries, such as the Humber, The Wash, the Thames, Severn

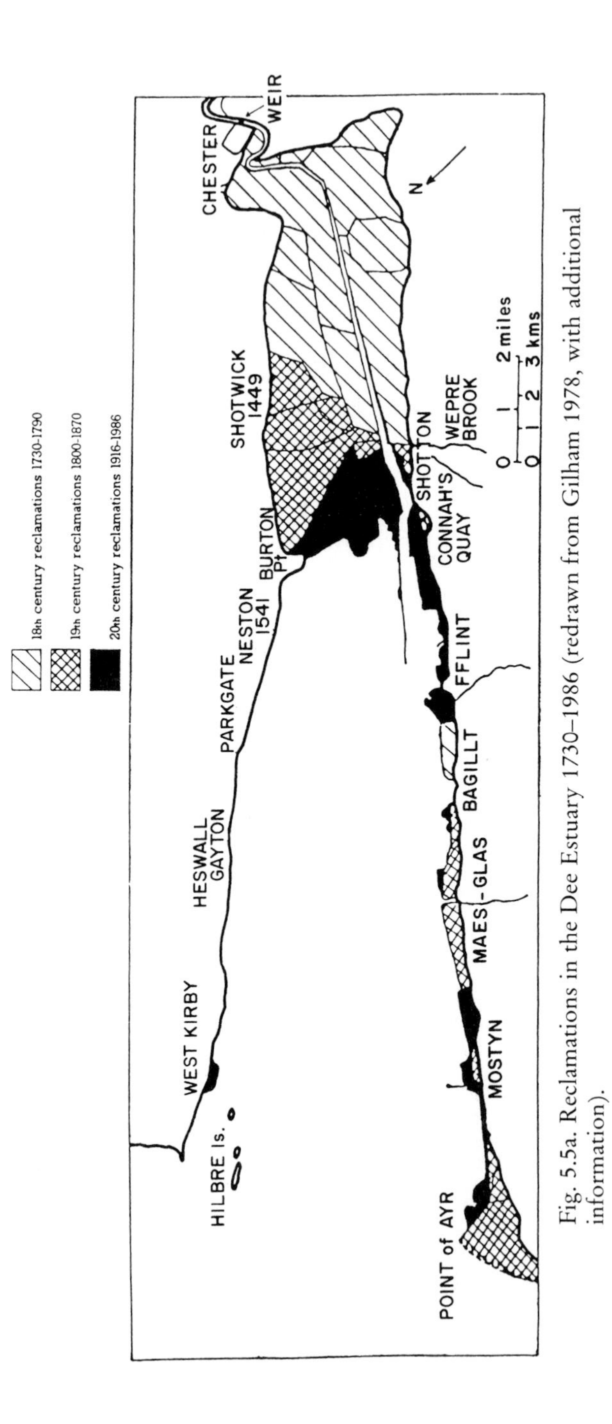

Fig. 5.5a. Reclamations in the Dee Estuary 1730–1986 (redrawn from Gilham 1978, with additional information).

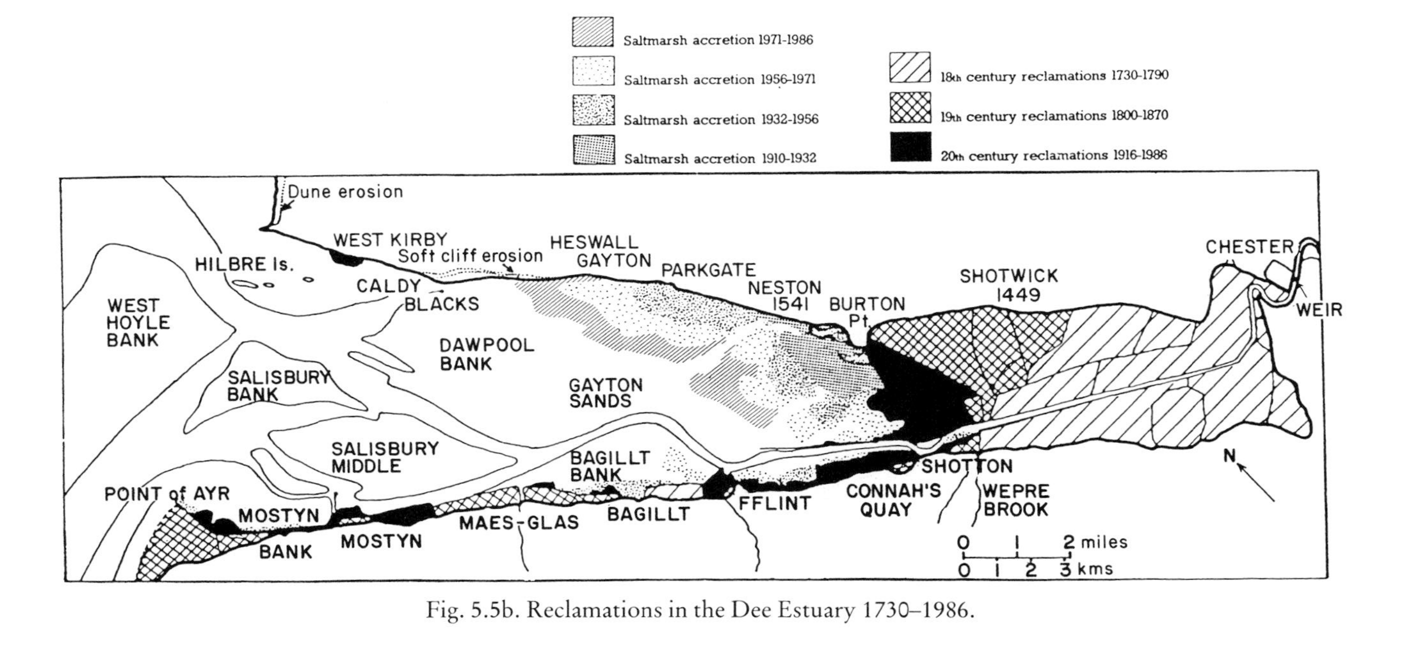

Fig. 5.5b. Reclamations in the Dee Estuary 1730–1986.

and the Dee. The immediate impact both on the saltmarsh itself and, in the longer term, on the functioning of the whole system was either not recognised or felt to be subservient to other needs. However, as the case of the Dee estuary illustrates, the knock-on effects of saltmarsh enclosures are perhaps only recognised, if at all, when it is too late.

Even the more sympathetic uses such as grazing, turf cutting, wildfowling or even bird watching did not always recognise the implications of 'reclamation'. Certainly in large sites like The Wash it was felt that saltmarsh enclosure did no long-term harm. New saltmarsh developed in front of the new sea wall and overall the nature conservation interest remained unaffected. However, it subsequently became clear that the picture was far more complex. Evidence from Admiralty charts and Ordnance Survey maps (Doody, 1987), studies of sediment budgets (Evans and Collins, 1987) and the position of low-water mark (Kestner, 1962) suggested that little accretion took place at the lower limits of the tidal flats. Hence, as saltmarsh enclosure proceeded over the last 100 years the intertidal area diminished as the foreshore steepened. This squeezing of the intertidal zone led the NCC at a public inquiry to oppose further reclamation at Gedney Drove End in recognition of the cumulative affects of enclosure and the threat to habitats and bird populations.

This oppostion was vindicated by subsequent decisions by the Local Authorities in relation to preparation of local plans which recommended a moratorium on reclamation. Today proposals for enclosure are infrequent and the NCC has embarked on the production of a nature conservation strategy for the whole site, which brings into the discussion all those with legitimate interests in the use of The Wash.

This case illustrates another important principle, which shows that scientific evidence may provide for a better understanding of the impact of man's activities on nature conservation interests (see Doody and Barnett, 1987) for a summary of research in The Wash). In its turn it has led to a reappraisal of an historically accepted practice. Much remains to be done in reconciling activities such as grazing management, military use and wildfowling, amongst others, but this has been an important part of the development of a nature conservation strategy at this site.

Unfortunately, the same acceptance of the need to stop treating tidal lands from the narrow perspective of the individual users and abusers is not seen elsewhere. In the NCC's Estuaries Review, 55 of the 155 identified estuary review sites have proposals for land claim. Whilst many of these are relatively small, involving rubbish and spoil disposal, they have to be considered against the background of an overall reduction in the area of tidal land of some 23% in historic times (Davidson *et al.*, in press).

In addition to this, other more major developments, notably associated with tidal energy barrages, pose a further significant threat to saltmarshes in sites like the Severn, Mersey, Humber and Duddon estuaries. The precise impact is not clear, though it is likely that the reduced levels of high tides and longer stand times at high water associated with energy generation needs, will cause massive changes to the saltmarsh zonation. Amenity barrages, such as that proposed in Cardiff Bay, will result in the destruction of the saltmarsh habitat as tidal waters are replaced by a freshwater lake. It is clear that, in the face of the competing demands on the saltmarsh habitat, their conservation can be achieved only by a change of attitude as far as exploitation is concerned. In addition, a more integrated approach to management, which takes account of the complexity of the environment in which the saltmarsh exists, will also be required.

Conclusion

The role of the conservationist lies in bringing together the competing demands and sectional interests and providing a framework in which best use can be made of the resource. As we have seen, this may involve manipulation of grazing management at one level, allowing wildfowling on a zoned basis at another, and the rehabilitation of saltmarsh as part of a sea defence strategy at yet another.

Maintaining or extending the present line of defence is a natural reaction when considering the protection of land and property from inundation from the sea. However, this has important consequences for the intertidal zone, notably saltmarshes, particularly in areas where sea level is rising relative to the land. In both Essex and Kent, where isostatic adjustment is resulting in a relative rise in sea level of up to 3 mm a year, the saltmarshes have been showing signs of erosion. The saltmarsh in the Dengie peninsula, for example, lost some 10% of its area between 1960 and 1981 (Harmsworth and Long, 1986). Over a longer timescale, Kirby (1985) estimates that in the Medway (North Kent) some 55% of the saltmarsh present in 1800 had been enclosed to the present day. During the same period a further 25% was lost through erosion and of the 20% remaining, this was considerably dissected and only a few per cent of the original area remains today.

From this perspective, and in the face of rising sea levels associated with global warming, erosion could extend beyond south and southeast England, if predictions are confirmed. If the current line of defence is maintained then the saltmarsh habitat will become an ever diminishing resource, squeezed between the land on the one hand and the rising sea on the other. Conservation in the face of this scenario will require an even more fundamental understanding of the development and recreation of saltmarsh, as a retreat at the coastal margin allowing new habitat to develop behind the sea-wall becomes an almost essential option, if we are to conserve the resource. This poses fundamental problems in relation to understanding the mechanisms associated with saltmarsh loss and predicting what changes will occur in the quality and quantity of new habitat created. It will also be necessary to consider the policy implications not only for the sea defence strategy but also agricultural and other current uses of the land which might be affected.

There are clearly many uncertainties both in trying to protect the remaining areas of saltmarsh from further destruction and in attempting to integrate their management and wise use within the wider environment. However, so long as their inherent value is recognised both for wildlife and as part of a functioning ecosystem, then decisions can be made which should ensure their better future. With so much saltmarsh within Sites of Special Scientific Interest the conservation interests have a pivotal role in contributing to unravelling and determining the complex issues involved.

References

ADAM, P. (1978) Geographical variation in British saltmarsh vegetation. *Journal of Ecology* **66**, 339–366.

ALLPORT, G., O'BRIEN, M. & CADBURY, C.J. (1986) *Survey of redshank and other breeding birds on saltmarshes in Britain, 1985*. RSPB Sandy, Report to NCC.

BARNES, F.A. & KING, K.A.M. (1953) The storm floods of the 1 February 1953. 2. The Lincolnshire coastline and the 1953 storm flood. *Geography* **38**, 141–160.

BEEFTINK, W.G. (1977) Saltmarshes. In R.S.K. Barnes (ed.) *The Coastline*. John Wiley and Sons, Chichester, 93–121.

BURD, F. (1989) *The Saltmarsh Survey of Great Britain. An Inventory of British Saltmarshes.* Research and Survey in Nature Conservation, **17**. Nature Conservancy Council, Peterborough.

CHAPMAN, V.J. (1938) Studies in saltmarsh ecology, Sections I to III. *Journal of Ecology* **26**, 144–179.

CHAPMAN, V.J. (1939) Studies in saltmarsh ecology, Sections IV and V. *Journal of Ecology* **27**, 160–201.

CHAPMAN, V.J. (1940) Studies in saltmarsh ecology, Sections VI and VII. Comparisons with marshes on the east coast of North America. *Journal of Ecology* **28**, 118–152.

CHAPMAN, V.J. (1941) Studies in saltmarsh ecology, Section VIII. *Journal of Ecology* **29**, 69–82.

CORKHILL, P. (1980) The *Spartina* problem. *Shooting Times and Country Life Magazine*, p. 29.

DALBY, R. (1957) Problems of land reclamation. 5. Saltmarsh in the Wash. *Agricultural Review* **2**, 31–37.

DAVIDSON, N.C., LAFFOLEY, D.A. & DOODY, J.P. (in press) Land-claim on British Estuaries: changing patterns and conservation implications. *Estuarine and Coastal Sciences Association Annual Meeting, University of Hull, September 1990.*

DAVIS, P. & MOSS, D. (1985) *Spartina* and waders in the Dyfi Estuary. In J.P. Doody (ed.) *Focus on Nature Conservation No. 5.* Nature Conservancy Council, Peterborough, 37–40.

DAVIES, M. (1987) Twite and other wintering passerines on the Wash saltmarshes. In J.P. Doody & B. Barnett (eds) *The Wash and its Environment. Research and Survey in Nature Conservation, No. 7.* Nature Conservancy Council, Peterborough, 123–132.

DOODY, J.P. (ed.) (1985) *Spartina anglica* in Great Britain. *Focus on Nature Conservation No. 5.* Nature Conservancy Council, Peterborough.

DOODY, J.P. (1987) The impact of reclamation on the natural environment of the Wash. In J.P. Doody & B. Barnett (eds). *The Wash and its Environment. Research and Survey in Nature Conservation, No. 7.* Nature Conservancy Council, Peterborough, 165–172.

DOODY, J.P. AND BARNETT, B. (1987) *The Wash and its Environment. Research and Survey in Nature Conservation, No. 7.* Nature Conservancy Council, Peterborough.

DOODY, J.P. (1990) *Spartina* – friend or foe? A conservation viewpoint. In A.J. Gray & P.E.M. Benham (eds) *Spartina anglica – A Research Review.* Institute of Terrestrial Ecology Publication No. 2. HMSO, London, 77–79.

EVANS, G. & COLLINS, M. (1987) Sediment supply and deposition in the Wash. In J.P. Doody & B. Barnett (eds). *The Wash and its Environment. Research and Survey in Nature Conservation, No. 7.* Nature Conservancy Council, Peterborough, 48–63.

EVANS, P.R. & PIENKOWSKI, M.W. (1984) Implications for coastal engineering projects of studies at the Tees Estuary on the effects of reclamation of intertidal land on shorebird populations. *Water Science and Technology* **16**, 347–354.

FULLER, A.J. (1982) *Bird Habitats in Britain.* T. & A.D. Poyser, Calton.

GILHAM R.M. (1978) *An Ecological Investigation of the Intertidal Benthic Invertebrates of the Dee Estuary.* Unpublished Ph.D. Thesis, University of Liverpool.

GOODMAN, P.J., BRAYBROOKS, E.M. & LAMBERT, J.M. (1959) Investigations into 'die-back' in *Spartina townsendii* agg. 1. The present status of *Spartina townsendii* in Britain, *Journal of Ecology* **47**, 651–677.

GRAY, A.J. (1972) The Ecology of Morecambe Bay V. The saltmarshes of Morecambe Bay. *Journal of Applied Ecology* **9**, 207–220.

GRAY, A.J. (1977) Reclaimed land. In R.S.K. Barnes (ed.) *The Coastline.* Wiley, Chichester, 253–270.

GRAY, A.J. & BENHAM, P.E.H. (1990) *Spartina anglica – A Research Review. Institute of Terrestrial Ecology Research Publication No. 2.* HMSO, London.

GRAY, A.J. & PEARSON, J.M. (1985) *Spartina* marshes in Poole Harbour, Dorset, with particular reference to Holes Bay. In J.P. Doody (ed.) *Spartina anglica in Great Britain. Focus on Nature Conservation, No. 5.* Nature Conservancy Council, Peterborough, 11–14.

GREEN, C. (1984) *Saltings and Sea Defences on the Gwent Levels.* Conference of River Engineers, Cranfield.

GREEN, R.E., JOHNSON, T.H. & COLLINS, D. (1984) *An Intensive Survey of Breeding Redshank on the Wash, 1982.* Unpublished Report, Royal Society for the Protection of Birds.

HARMSWORTH, G.C. & LONG, S.P. (1986) An assessment of saltmarsh erosion in Essex, England, with reference to the Dengie peninsula. *Biological Conservation* **35**, 377–387.

HAYNES, F.N. & COULSON, M.G. (1982) The decline of *Spartina* in Langstone Harbour, Hampshire. *Proceedings of the Hampshire Field and Classical Archaeological Society* **38**, 5–18.

HILL, M. (1987) Vegetation Change in Communities of *Spartina anglica* (C.E. Hubbard) in Saltmarshes of North-west England. *Contract Surveys, No. 9.* Nature Conservancy Council, Peterborough.

HILL, M. (1988) *Saltmarsh Vegetation of The Wash. An Assessment of Change From 1971–1985. Research and Survey in Nature Conservation, No. 13.* Nature Conservancy Council, Peterborough.

HOLDER, C. & BURD, F. (1990) *Saltmarsh Restoration Sites in Essex. Contract Surveys, No. 83.* Nature Conservancy Council, Peterborough.

HUBBARD, J.C.E. & STEBBINGS, R.E. (1967) Distribution dates of origin and acreage of *spartina townsendii* (s.l.) marshes in Great Britain. *Proceedings of the Botanical Society of the British Isles* **7**, 1–7.

KESTNER F.J.T. (1962) The old coastline of The Wash – a contribution to the understanding of loose boundary processes. *Geographical Journal* **128**, 457–478.

KIRBY, R. (1985) The recent history of the lower Medway saltmarshes. In J.P. Doody (ed.) *Spartina anglica in Great Britain. Focus in Nature Conservation, No. 5.* Nature Conservancy Council, Peterborough, 18–20.

LONG, S.P. & MASON, C.F. (1983) *Saltmarsh Ecology*. Blackie, Glasgow.

MACEY, M.A. (1974) Repord 1d. Survey of semi-natural reclaimed marshes. In *Aspects of the Ecology of the Coastal Area in the Outer Thames Estuary and the Impact of the Proposed Maplin Airport.* Report to Department of the Environment by the Natural Environment Research Council.

MARKER, M.E. (1967) The Dee Estuary: its progressive silting and saltmarsh development. *Transactions of the Institute of British Geographers* **241**, 65–71.

RATCLIFFE, D.A. (1977) *A Nature Conservation Review – The Selection of Biological Sites of National Importance for Nature Conservation.* Cambridge University Press, Cambridge.

RANDERSON, P.F. (1984) *Saltings and Coastal Stability – A Biologist's View.* Conference of River Engineers, Cranfield.

RANWELL, D.S. (1972) *Ecology of Saltmarshes and Sand Dunes.* Chapman and Hall, London.

RANWELL, D.S. (1964) *Spartina* saltmarshes in southern England (III). Rates of establishment, succession and nutrient supply at Bridgwater Bay, Somerset. *Journal of Ecology* **52**, 95–105.

RANWELL, D.S. (1974) The saltmarsh to tidal woodland transition. *Hydrobiological Bulletin* **8**, 139–151.

RODWELL, J. (in press) *British Plant Communities*, Vol. 4. Cambridge University Press, Cambridge.

ROYAL COMMISSION ON COASTAL EROSION AND AFFORESTATION (1911) *Third (and Final) Report Vol. III, Part V., The Reclamation of Tidal Lands.*

STRICKNEY, W.H. (1908) *Plan Showing Reclamation from the River Humber, Holderness.* Appendix XXX to Evidence to the Royal Commission on Coastal Erosion and Afforestation, Second Report.

STUBBS, A. (1982) *A Preliminary Account of Saltmarshes in Great Britain.* Appendix B. The invertebrate fauna of saltmarsh. Chief Scientist Team Note No. 31.

THORNTON, D. & KITE, D.J. (1990) *Changes in the Extent of the Thames Estuary Grazing Marshes* Internal Report, Nature Conservancy Council, London.

TRUSCOTT, A. (1985) Control of *Spartina anglica* on the amenity beaches of Southport In J.P. Doody (ed.) *Spartina anglica in Great Britain. Focus on Nature Conservation, No. 5.* Nature Conservancy Council, Peterborough, 64–69.

TUBBS, C.R. (1981) Current and future planning and conservation. In F. Stranack & J. Coughlan (eds) *Solent Saltmarsh Symposium.* Winchester, 25–30.

WHITE, D.A. (1982) *Dee Estuary Vegetation Monitoring 1971–1979.* Wales Field Survey Unit Report. Nature Conservancy Council, North Wales Region.

WHITESIDE, M.R. (1987) *Spartina* colonisation. In N.A. Robinson & A.W. Pringle (eds) *Morecambe Bay, an Assessment of Current Knowledge.* Resource paper for Centre for N.W. Regional Studies and Morecambe Bay Study Group, 118–129.

6

Engineering significance of British saltmarshes

A.H. BRAMPTON

Introduction

Saltmarshes are found along a substantial part of the UK coastline, amounting to some 20% in England and Wales, mainly in or close to estuaries. The richness of the flora and fauna of these areas has resulted in 80% of saltmarshes being designated as Sites of Special Scientific Interest (SSSI).

Saltmarshes have therefore attracted a disproportionately high level of interest from biologists and conservationists compared with other types of coast. In direct contrast, saltmarshes have evoked very much less interest from coastal engineers than would have been expected given the percentage of shoreline they occupy.

This difference in interest is not as surprising as it may appear at first sight. Flat, alluvial coasts with saltmarshes have never been considered attractive locations for residential or commercial development. Occasionally there is some industrial development in the vicinity of estuarine ports, but normally the area landward of saltmarshes is given over to agriculture. This limits the benefits which have to be calculated before, say, a coast protection scheme is undertaken.

It is also worth pointing out that much of the agricultural land protected by saltmarshes has been 'won' from the sea in the past, by gradually reclaiming the upper portion of those saltmarshes. In many areas the historical tendency of accretion of saltmarshes continues. In other areas, even if the situation has deteriorated, the saltmarshes are in sheltered locations and simple clay banks are all that may be needed to forestall flooding of the hinterland. In view of all this it is perhaps not surprising to find that standard coastal engineering textbooks have little or nothing dealing with the management or repair of saltmarshes.

In recent times, however, there has been a widespread and serious erosion of saltmarshes around the coast of the UK, particularly in Essex, Hampshire and along the shores of the Severn Estuary. This has been sufficiently serious to force a revision of the attitudes of coastal managers. This paper reviews the engineering significance of saltmarshes, and the problems caused by their erosion.

Saltmarshes as a coastal defence

The greatest significance of saltmarshes to a coastal manager lies in their ability to dissipate wave energy so that little, if any, remains at their landward limit. Given this situation, a modest clay embankment will protect the hinterland from flooding during exceptionally high tides. If the saltmarshes erode, allowing larger waves to reach such an embankment, then it may become necessary to strengthen and raise the embankment to ensure its survival and to prevent overtopping. Even a modest embankment may cost £1000 per metre run and, given that about 2000 km of UK shoreline is protected by saltmarshes (Doody 1991), their value as coast defences is clearly very considerable.

The efficiency of a healthy saltmarsh as a coastal protection was investigated by Hydraulics Research (1980) in a study of defences along the Severn Estuary, carried out jointly for Severn-Trent Water Authority, Wessex Water Authority and CEGB. Figures 6.1 and 6.2 (reproduced from Owen, 1984) show a typical saltmarsh/ embankment cross-section and an example of the way that wave heights travelling towards the embankment are reduced by breaking and shoaling. The calculations shown in Fig. 6.2 are no more than speculative examples; unfortunately, there is very little information from prototype saltmarshes to verify numerical or scale physical models, and this shortcoming is discussed later in the paper.

The scale physical model experiments carried out in 1980 and subsequently still provide the best quantitative advice to coastal engineers concerned with providing adequate defences on a saltmarsh coast. The efficiency of a high, wide saltmarsh acting as a natural berm in front of a sea wall is illustrated in Figs 6.3 and 6.4 which are also from Owen (1984). These show how a reduction in saltmarsh width will cause an increase in the volume of water overtopping the defences, or require an increase in the crest height of the wall to provide the same standard of protection. Similar calculations, not illustrated, can be carried out to show the consequences of a fall in saltmarsh levels.

A particularly striking example of the efficiency of saltmarshes can

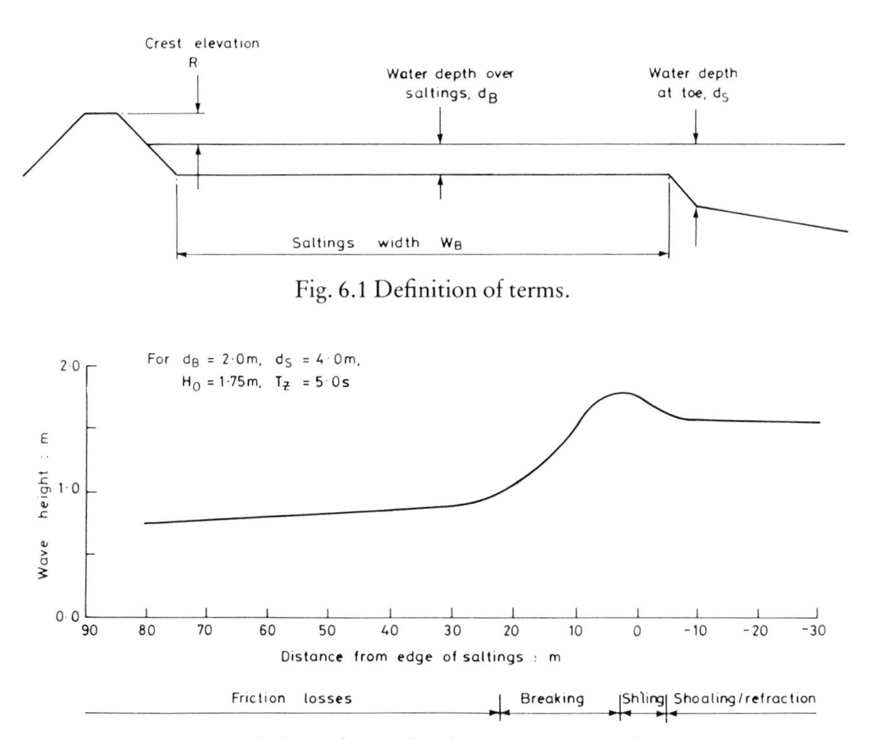

Fig. 6.1 Definition of terms.

Fig. 6.2 Estimated wave height variation over saltings.

be found on the South Wales coastline. The Gwent Levels, stretching eastwards from Cardiff towards Chepstow, have two types of defence. Where there are healthy saltings, the main defence is grassed earth embankment with a crest level ranging from about 8.8 to 9.5 m AOD. Where there are no saltings, the defence levels are at about 10.5 m AOD, and comprise a near vertical wall capped by a wave return wall and fronted by a rock slope up to 8 m high (Green, 1984). There is no conflict between conservationists, coastal engineers and the Treasury on which of the two types of defence is preferable.

Saltings and estuarial stability

It is normal in estuaries for saltmarshes to occupy a large percentage of the intertidal zone. A sudden erosion of saltmarshes, for example as a result of disease in the vegetation, could therefore result in a substantial increase in the volume of water flowing into and out of the estuary during a tidal cycle. At the same time, of course, the

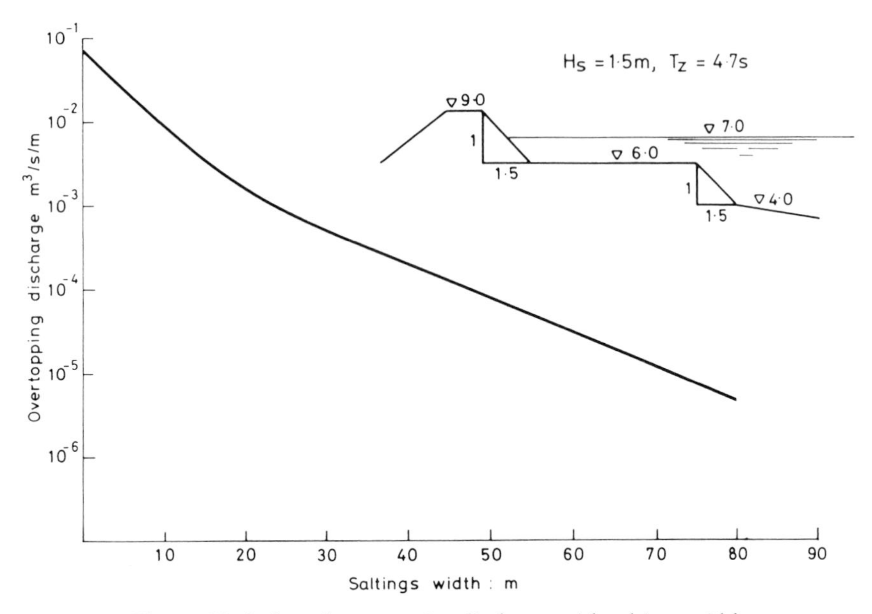

Fig. 6.3 Variation of overtopping discharge with saltings width.

cross-sectional area of the channels will also be enlarged. It is not possible to give any general guidelines on the resulting changes in the morphology of an estuary. The tidal currents will almost certainly increase in some areas, however, thus causing further erosion of the mud and any local areas of saltmarsh, adding to the changes in the flow regime. Conceivably an estuary may become completely destabilised by such a sequence of events. It is more likely, however, that a new state of dynamic equilibrium will be established. Unfortunately, predicting how an estuary would alter in response to saltmarsh losses is not possible, except perhaps in a very general way. Since many estuaries contain commercial ports or marinas with fixed facilities, the financial consequences of rapid morphological changes are substantial. In some cases, the result may be an increase in the rate of dredging to maintain adequate depths in approach channels and at berths or quays. In others, the loss of saltmarshes may result in unacceptable wave action within previously sheltered areas, necessitating installation of breakwaters or similar structures. Faced with such consequences, and a high degree of uncertainty regarding the extent and pace of changes, maintaining an existing estuarial regime is usually regarded as the safest commercial option.

Whilst considering estuarial stability, the other side of the question also needs to be considered, that is, the problems arising from an

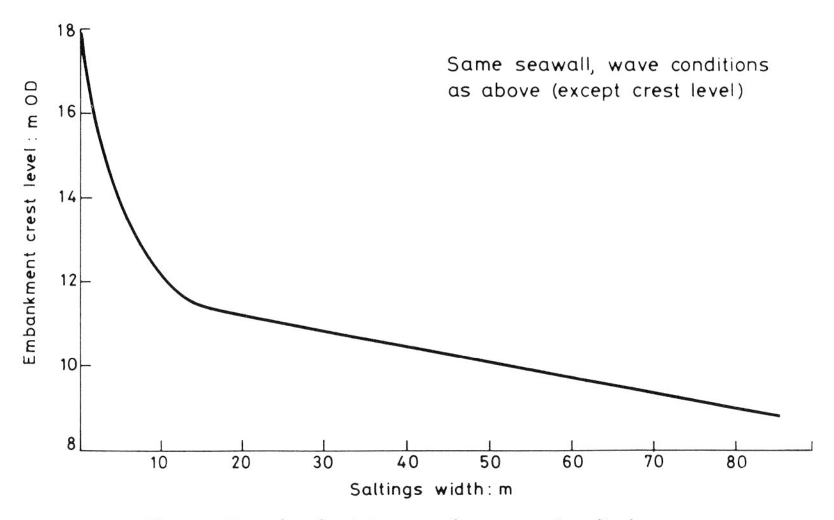

Fig. 6.4 Crest levels giving equal overtopping discharge.

increase in saltmarsh area. This situation may arise as a result of a general siltation of an estuary (as in certain west coast sites in the UK) or perhaps because of the spread of more successful vegetation (for example the arrival of *Spartina anglica*). The estuarial regime will be altered both by the reduction in tidal volume, and in the cross-sectional area of channels, with a greater risk of siltation in some areas. Again there are obvious financial implications for commercial ports, as a result of increased dredging.

Engineering solutions to saltmarsh problems

Over the last 20 years, engineers have been employed to design and construct schemes to counter problems on saltmarsh coasts, usually arising from rapid and widespread erosion of the saltmarshes. Most commonly, assistance has been sought by the National Rivers Authority (previously the Regional Water Authorities) who have the responsibility for protecting low-lying land close to the sea against inundation. In the UK, embankments, walls and other structures built to fulfil this role are referred to as 'sea defences', whilst 'coast protection' covers often very similar structures built by local authorities (ie District Councils) to protect urban frontages, particularly against erosion of the land.

Apart from the National Rivers Authority, port and marina

operators, and private landowners also occasionally request assistance as a result of the erosion of saltmarshes. In virtually all cases, the remit of the engineer is to produce a solution to counter an increased threat of wave action and/or overtopping at the rear of the saltmarsh, at reasonable if not minimum cost. Often assistance is not requested until the danger of flooding is acute, or flooding has actually occurred. In such circumstances it is feasible only to treat the results of saltmarsh erosion rather than the underlying causes. Usually the response is to strengthen the main line of defence, by raising the crest level and armouring the front face of the wall. Engineers are fully aware of the dangers inherent in the 'fire brigade' approach to coastal defences; after all, this is the approach often used on other types of coast.

Against this background, it is encouraging to report a growing awareness of the advantages of a management system for sea defences. This should allow problems to be identified at an earlier stage, and hence permit a wider range of remedial measures to be considered. The example set by the National Rivers Authority (Anglian Region) in embarking upon both a major Sea Defence Managements Study and the active management of saltmarshes in Essex deserves special mention.

In the long term, the most effective solutions to coastal erosion problems are those which are sympathetic to the natural processes of sediment transport. They also tend to be more acceptable in terms of their environmental impact. This is likely to be the case with saltmarshes and muddy intertidal shores, just as much as with other beaches.

Unlike beaches of sand or shingle, however, there is very much less information on sediment transport or on changes in level on salt-marsh coasts. The difficulty of safe access to the intertidal zone, the need to carry out full three-dimensional surveys rather than cross sections, and the problems of accurately measuring levels are all contributing factors to the lack of quantitative data on levels. Without this information, however, it becomes impossible to calculate a sediment 'budget' for such a coast. In turn, this obscures the long-term transport patterns and it is not possible to determine what processes have to be countered to prevent further erosion. As a result it is difficult to see how engineers could design or predict the performance of, say, a nourishment scheme. It may even be difficult to monitor its effects, and this could prevent the use of one very powerful weapon of coastal management.

The vital role of vegetation in maintaining saltmarshes is also a complicating factor in devising appropriate remedial measures. As with sand dunes, it can be expected that a single severe winter, or several in succession, will cause damage which cannot be repaired in a single 'growing' season. Genuine long-term trends may therefore be difficult if not impossible to determine within a few years of monitoring. It is also true that methods of management which involve a modest input of manpower year after year are less easy to design and control than a short intensive work campaign. In most cases the civil engineering industry is accustomed to the latter kind of work. On saltmarshes, as with sand dunes, the best management methods are likely to be more like grounds maintenance than a design and build contract. The establishment of 'polders' on the Essex coast is a good example of the type of work that is likely to be required (different management methods are reviewed in two reports by Hydraulics Research Ltd. 1987, 1988).

Future research requirements

On saltmarsh coasts especially, any major intervention works to prevent flooding or reduce siltation are likely to be examined with keen interest by conservation bodies. Some form of environmental impact assessment is likely to be demanded, and this requires a clear idea of the morphodynamic and hydraulic changes likely to occur following the proposed works. This is already standard procedure for other types of coast, whether the works are structures such as sea walls, groynes or breakwaters, or comprise active beach management techniques such as nourishment or recycling.

The coastal management fraternity must therefore turn its attention to improving the understanding of saltmarshes, and of intertidal mudflats, to be able to predict with confidence the consequences of any interference in those areas. It is worth making the point here that such assessments should work in two directions. It is entirely reasonable to ask that a proposed sea wall designed to prevent flooding should be examined on the basis of possibly harming the biological interests of a saltmarsh. Equally, work by conservation bodies to limit saltmarsh areas in favour of mudflats (which are more useful to wading birds) should be subject to similar examination on the basis of possible impacts on sea defences.

It therefore seems in everybody's interest to improve knowledge of the hydraulic and morphodynamic behaviour of saltmarshes (and intertidal mudflats). Monitoring of levels within the intertidal zone seems a vital area for effort. Without such information from the existing coast, making predictions, of future conditions is quite impossible. Similarly, information is required on water movements over mudflats and saltmarshes. It seems likely, for example, that waves are more effectivly dissipated if mud is covered by vegetation than if it is not. Confirmation and measurements of such processes are badly needed if appropriate methods of managing such coasts are to be developed.

Acknowledgements

The author wishes to acknowledge the contribution of numerous colleagues at Hydraulics Research on whose work he has drawn to produce this paper, particularly Mr Michael Owen and Mr Jack Welsby. Financial assistance for research into the role of saltmarshes in coastal defence was provided by the UK Ministry of Agriculture, Fisheries and Food; the views expressed in this paper are, however, those of the author and not necessarily of the Ministry.

References

DOODY, J.P. (1991) *Sea defences and nature conservation: threat or opportunity.* Institute of Water and Environmental Management, River Engineering Section, Winter Meeting (1 February, 1991).

GREEN, C. (1984) *Saltings and sea defence on the Gwent Levels* MAFF Conference of River Engineers, Cranfield.

HYDRAULICS RESEARCH LTD. (1980) *Design of sea walls allowing for wave overtopping.* Report EX 924 Hydraulics Research Ltd, Wallingford.

HYDRAULICS RESEARCH LTD. (1987) *The Effectiveness of Saltings.* Report SR 109, Hydraulics Research Ltd., Wallingford.

HYDRAULICS RESEARCH LTD. (1988) *Review of the Use of Saltings in Coastal Defence.* Report SR 170, Hydraulics Research Ltd, Wallingford.

OWEN, M.W. (1984) Effectiveness of saltings in coastal defence, MAFF Conference of River Engineers, Cranfield.

7

Tidally influenced marshes in the Severn Estuary, southwest Britain

J.R.L. ALLEN

Introduction

Sediments formed largely on tidally influenced marshes and high tidal mudflats have accumulated over postglacial (Flandrian) times on the margins of the Severn Estuary and inner Bristol Channel to a typical thickness of 10–15 m and over an outcrop today measuring some 840 km^2. According to circumstances, the marshes varied from minerogenic, when and where the supply of tidal silt exceeded the availability of organic sedimentary material provided locally by marsh plants, to organogenic with a relative decline in tidal supplies. Most of the outcrop of estuarine alluvium has now been reclaimed, but active tidally influenced marshes can still be seen in the area (Smith, 1979) and prove to be wholly minerogenic in character. Organogenic marshes, however, recorded as peats and organic-rich silts, were developed on an estuary-wide scale on several occasions during a mid-Flandrian epoch dating from *c.* 6500–2500 conventional radiocarbon years ago.

Although of only limited extent, the active marshes (1.5 km^2) on the borders of the Severn Estuary and inner Bristol Channel afford an important example in western Britain of a coastal environment of significance to sedimentologists, geomorphologists and biologists. The marshes are also significant for engineers, on account of their contribution to coastal defence. What is of particular concern to all is the as yet incompletely understood but substantial historical instability of the high-water shoreline in the area, along which the mudflats and marshes lie.

Setting

The Severn Estuary and inner Bristol Channel in southwest Britain constitute a partly rock-bound system (Fig. 7.1a, b), in which the

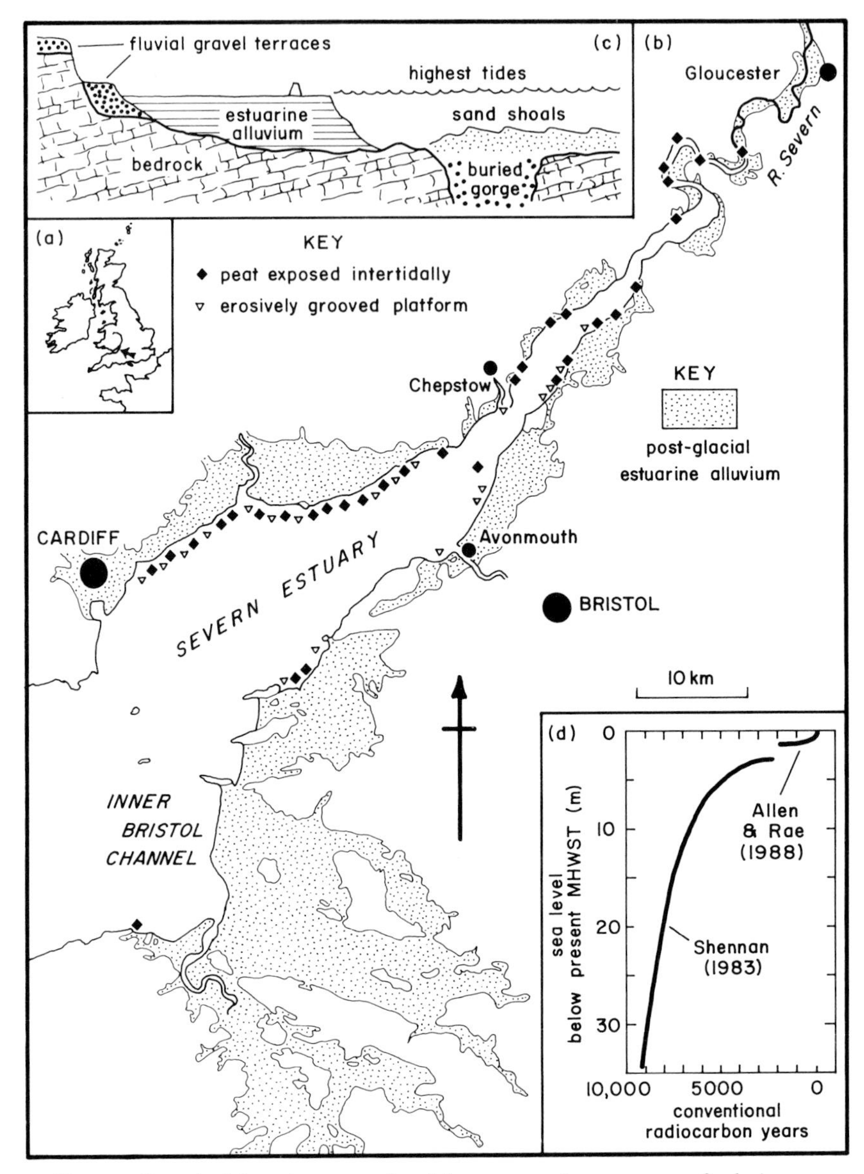

Fig. 7.1. Postglacial geology, erosional features, and movement of relative sea level in the inner Bristol Channel and Severn Estuary.

post-glacial estuarine alluvium (Fig. 7.1c) incompletely mantles a comparatively level outer bedrock valley, of mild but intricate relief into which rivers have cut deep and now largely buried gorges (Codrington, 1898; Hawkins, 1962; Anderson and Blundell, 1965, Anderson, 1968, 1974; Williams, 1968; Whittaker and Green, 1983).

On the margins of the outer valley, and locally on its floor, occur fluvial terrace gravels of late Pleistocene date (Wills, 1938). The evolution of the bedrock surface is poorly understood, but at different times involved the operation of fluvial (Wills, 1938), marine (Andrews *at al.*, 1984), and periglacial (Allen, 1987a) processes.

The Severn Estuary and inner Bristol Channel open southwestward and then westward toward the Celtic Sea and the prevailing winds (Fig. 7.1a). Their regime is strongly macrotidal, the astronomical tides at Avonmouth (Hydrographer of the Navy, 1990) having an extreme range of 14.8 m, mean spring and mean neap ranges of respectively 12.3 m and 6.5 m, mean high-water springs rising to 13.2 m above tidal datum, and mean high-water neaps of 10.0 m. The mean high-water of spring tides, which is approximately the level of the active tidal wetlands, attains an altitude of about 6 m above Ordnance Datum (OD) in the inner Bristol Channel, rising to a maximum of about 8.5 m OD some 10 km downstream from Gloucester at the head of the inner Estuary. The tidal currents are correspondingly vigorous (Hamilton, 1979; Crickmore, 1982; Uncles, 1982, 1984; Stephens, 1986; Uncles *at al.*, 1986) and, because of the size and aspect of the system, substantial storm surges (Lennon, 1963a, 1963b) and powerful waves (Shuttler, 1982; Smith, 1983; Cook and Prior, 1987) can at times arise. Fine sediment is introduced mainly by the rivers, which supply some 10^6 t a^{-1} and the water body holds the equivalent of the order of 10–20 years of fluvial supply (Collins, 1983; Allen, 1990a, 1992).

Postglacial movements of sea level

The tidally influenced marshes and high tidal mudflats of the Severn Estuary and inner Bristol Channel formed against a background of a generally rising 'sea level (long-term smoothed trend), at first rapid but afterwards increasingly gradual. This general trend (Fig. 7.1d), using mean high water of spring tides as the reference level (Shennan, 1983), is clear up to about 2500 conventional radiocarbon years ago from 311 radiocarbon-dated sea-level index points assembled by Heyworth and Kidson (1982). The later trend (a minimum curve) is based on 45 measurements of saltmarsh accretion rate spread over the region, with the dating provided by archaeological and geochemical evidence (Allen and Rae, 1988). Using documentary sources in addition, Allen (1991) obtained a closely similar curve for the inner Estuary based on 94 accretion measurements.

The long-term rate of rise in the inner Bristol Channel and Severn Estuary decreases from the order of 10–20 mm a^{-1} in the early Flandrian to a few mm a^{-1} during the earlier part of the mid-Flandrian epoch above, falling further to the order of 1 mm a^{-1} in recent millenia. Allen and Rae's (1988) and Allen's (1991) data point to an acceleration in the rate of rise of the level of mean high-water springs over the past few centuries to a current value of 3–5 mm a^{-1}, possibly because of reclamation-related or other changes of tidal regime (e.g. Woodworth, *at al.*, 1991), but there is no evidence from tide-gauge records that mean sea level is behaving similarly (Woodworth, 1987, 1990). The fact that several peats and other organic beds formed widely in the area during the mid-Flandrian epoch suggests that fluctuations of sea level (see also Shennan, 1986a, 1986b), with a period of the order of 500–1,000 years, but as yet uncertain amplitude, may lie concealed within the smooth trends sketched in Fig. 7.1d. According to Shennan (1989a), a downward crustal movement of less than 1 mm a^{-1} is occuring in the area.

Standard postglacial stratigraphic sequence

A voluminous local literature (review in Allen, 1990a), and a substantial number of mainly unpublished commercial boreholes, make it clear that the post-glacial estuarine alluvium, of predominantly mudflat-marsh origin, presents a remarkably uniform general sequence throughout the inner Bristol Channel and Severn Estuary (Fig. 7.2). Excluding the incised river valleys, where a variety of local sequences are developed, the main part of the succession is composed of the Wentlooge Formation (Allen, 1987b; Allen and Rae, 1987), informally divisible broadly into three. The lower Wentlooge Formation begins variously with thin gravels, sands, organic-rich palaeosols and rooted peats. These grade up into silty sands and then thick bluish grey to greenish grey sandy to clayey silts. The middle unit, ranging from below to several metres upward from OD according to location in the area, consists of interleaved peats, organic-rich beds and sandy to clayey silts. Thickening and merging toward the inland margins of the alluvial outcrop, the peats are rooted and represent a wide range of facies (woodland, carr, reedswamp, raised bog) (e.g. Locke, 1971; Heyworth and Kidson, 1976; Coles and Coles, 1986; Smith and Morgan, 1989). The various deposits of the middle unit are particularly rich in animal (including human) footprints and in

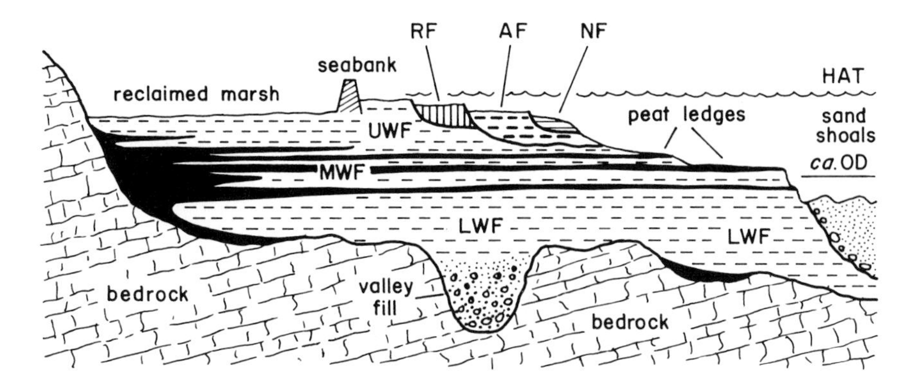

Fig. 7.2. 'Standard' postglacial geological sequence in the inner Bristol Channel and Severn Estuary.

evidence for prehistoric human activities based on minerogenic as well as organogenic marsh sites ranging from temporary encampments to permanent settlements (e.g. Coles and Coles, 1986; Green, 1989; Whittle *at al.*, 1989). A return to bluish to greenish grey and then green-brown mottled silts is seen in the upper Wentlooge Formation. At all but a very few locations, the continuing accumulation of the Wentlooge Formation has been halted because of marsh reclamation. Three other silt deposits – the Rumney Formation, the Awre Formation and the Northwick Formation – have accumulated in the area over the last few centuries and represent the active marshes.

A major effect of rising sea level in the area is that the various sedimentary environments of the Severn Estuary are being forced bodily northeastward up the Severn Vale, with the postglacial fine sediments in particular experiencing a process of reworking describable as 'stratigraphic roll-over' (Allen, 1990a). Thus today the mid-Flandrian peats are widely exposed on cliffs and ledges in the intertidal zone, and grooved platforms of considerable extent can be seen to have been scoured into the Wentlooge Formation silts (Fig. 7.1). As Allen (1990a) noted, the sands of mid and low intertidal shoals in the contemporary estuary are erosively juxtaposed against the lower beds of the Wentlooge Formation.

Older marshes

Little is undersood of the nature of the sub-fossil tidally influenced marshes preserved within the Wentlooge Formation. The potential of the exposed Wentlooge Formation for sedimentological and palaeogeomorphological studies has yet to be realised, and the floristic work attempted to date has undoubtedly laid undue emphasis on thick peat sequences and sites well inland (Jefferies *at al.*, 1968; Coles and Coles, 1986; Smith and Morgan, 1989; *cf.* Seddon, 1964).

Preliminary work, however, indicates that the sub-fossil marshes may have been comparatively simple morphologically. Stratigraphic relationships evident in air photographs of the coast, and the exposures on the ground, together with the geometry of the basal contacts of peat beds (regressive overlap of Shennan, 1986b), hint in the main at a wetland landscape in which a small number of relatively widely spaced large tidal channels were associated with comparatively smaller gullies and rills set in smooth areas of considerable extent. Whether it is a matter of the extreme tidal range or the considerable muddiness of the system, there is little in these features to suggest the dense and deep creek networks of the sandy Morecambe Bay (Gray, 1972; Gray and Scott, 1987) and Solway (Marshall, 1962) marshes and those of the Norfolk coast (Steers, 1960; Funnell and Pearson, 1989). The replacement of an organogenic marsh by a minerogenic one during the accumulation of the Wentlooge Formation, giving rise to a transgressive overlap (Shennan, 1986b), seems almost invariably to have been accompanied by the substantial erosional widening (but eventual infilling) of the larger channels and the cutting of new ones into previously smooth wetland areas (Fig. 7.3). Invariably these features sharply truncate the surrounding bedding, and their margins at many places reveal commonly spectacular evidence for large-scale mass movements, in the form of deeply ranging, curved slip surfaces and back-tilted peat beds.

Marsh reclamation

Allen and Rae's (1987) broad division of the reclaimed wetland areas between a Wentlooge Surface, considered to represent Romano-British intakes, and a ridge-and-furrowed (Hall, 1981, 1982; Astill, 1988) Oldbury Surface, supposedly recording medieval reclamation, is a good indication of differences in early agricultural practices, but

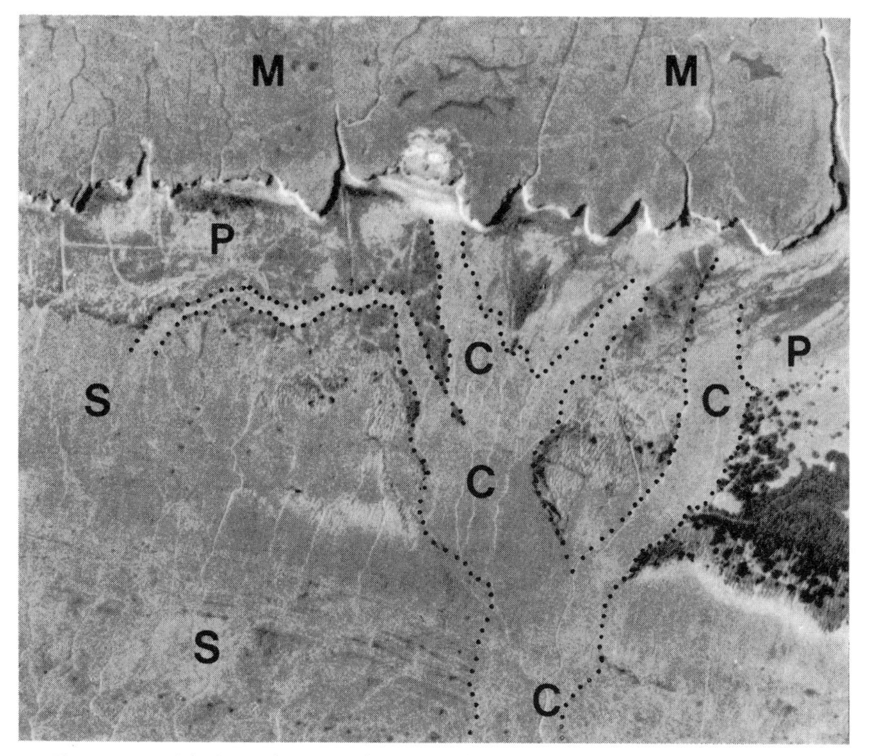

Fig. 7.3. Aerial view of traces of erosive, discordantly filled creeks (C, dotted boundary) cutting peat ledge (P) and stratigraphically lower silts (S), overlooked by contemporary saltmarsh (M), Rumney Great Wharf, Gwent. Bases of Romano-British drains visible on peat ledge in upper left. Area measures approximately 500 × 450 m. Crown copyright reserved.

must now be seen as an unreliable guide to the timing of the construction of sea defences along the Severn Estuary and inner Bristol Channel. Increasingly, detailed local archaeological work, especially field walking, is proving the Romano-British origin of the Oldbury Surface. Medieval and post-medieval reclamation was on a comparatively minor scale compared with this early activity. A general summary appears in Fig. 7.4.

Consider first the estuarine alluvium on the right bank of the system. The Wentlooge Level, between Cardiff and Newport, was almost certainly reclaimed in the Roman period (Allen and Fulford, 1986; Allen, 1990b) and it seems likely, judging from the distribution of archaeological sites (Fig. 7.4), that much of the Caldicot Level ranging northeastward to the Wye at Chepstow was also then taken in. The New Grounds at Lydney is a post-medieval intake, but the

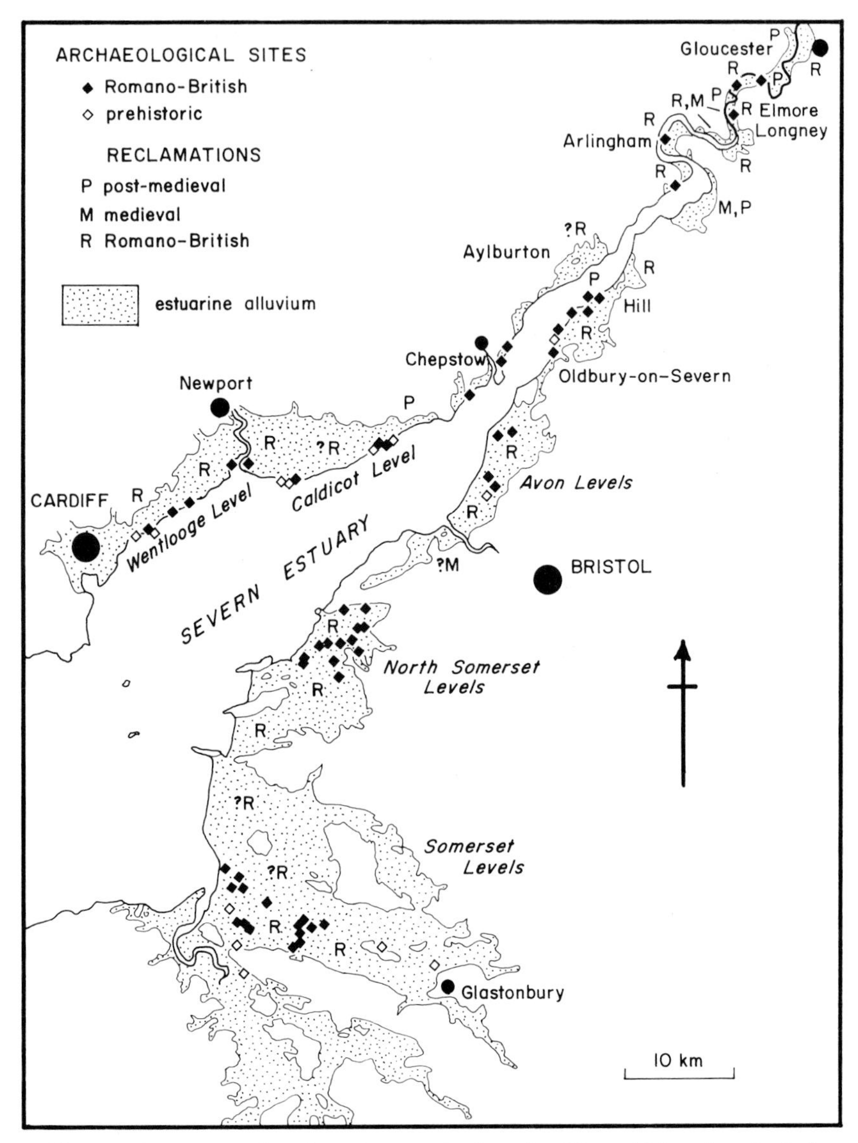

Fig. 7.4. Summary of reclamations and of wetland/intertidal archaeological (occupation) sites in the inner Bristol Channel and Severn Estuary.

proximity of the Aylburton Roman villa (Allen and Fulford, 1988) hints that the adjoining Oldbury Surface may record Romano-British reclamation. The small intakes of this early date upstream at Awre and Rodley are outweighed in the inner Estuary by medieval and later ones (Allen and Fulford, 1991).

The left bank (Fig. 7.4) reveals substantial Romano-British reclamations at Elmore (Allen and Fulford, 1990), Longney (Allen and Fulford, 1991), Moreton Valence (Allen and Fulford, 1991), and Arlingham (Allen 1990c). There is a small medieval reclamation at Longney (Allen and Fulford, 1991) and, followed by substantial post-medieval ones, at Slimbridge (Allen, 1986). Dated settlement sites on the alluvium have established the Romano-British age of the large intake at Hill and Oldbury-on-Severn (review in Allen, 1990a). Romano-British pottery, together with other evidence for occupation, including a substantial villa, is widely recorded from the Avon Levels, and the North Somerset Levels (review in Allen, 1990a). There has been little subsequent reclamation in these areas. The status of the main Somerset Levels in detail is far from clear. Much Romano-British activity occured in the area (Cunliffe, 1966; McDonnell, 1985, 1986), which could have been promoted by natural as well as artificial sea defences, but the size and complexity of these Levels makes it highly likely that the degree of interest in wetland reclamation and drainage varied with time and local circumstances, as is certainly evident from medieval times onward (Williams, 1970).

The character of these wetland environments at the time they were reclaimed and thus fossilised has not entirely been obliterated by nearly two millenia of varied human use. The narrower marshes appear to have sloped gently downward toward their inland margins at a gradient of the order of 10^{-3} m m^{-1}. The broader ones, however, are much more level and also uneven. Here and there, as at Arlingham (Allen, 1990c) and the Wentlooge Level (Allen, 1990b), the survival of drains of sinuous plan behind the sea defences points to the exploitation by the reclaimers of pre-existing meandering tidal creeks. Especially in the inner Estuary (e.g. Allen and Fulford, 1990), the occurence of some creeks and gullies on the marshes prior to reclamation is suggested by curvilinear features elevated above their surroundings and composed of sediments slightly coarser than the norm (Fig. 7.5). However, there is no suggestion in this evidence that the wetlands at reclamation, any more than the older marshes (see above), carried the dense arrays of deep creeks and gullies seen on many other British saltmarshes. As with river flood plains (Allen, 1985a; James, 1986; Pizzuto, 1987; Knight, 1989), to which tidally influenced marshes are morphologically and hydraulically analogous, the (minerogenic) marsh sediments along the Severn Estuary and inner Bristol Channel decrease in grain size (e.g. Allen and Fulford, 1990) toward their inland margins (Fig. 7.5). The commonly observed

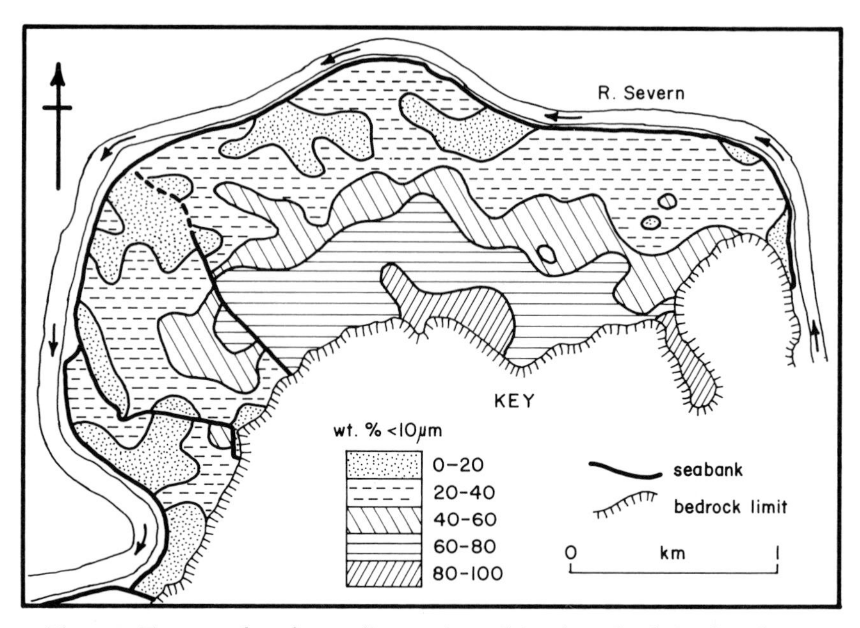

Fig. 7.5. Texture of surface sediments in reclaimed wetlands in the Elmore area, 5–10 km downstream from Gloucester.

downward surface dip also in this direction points to an inland decrease in the longer-term deposition rate, as is to be expected where diffusion partly determined the sediment supply. Convection may have dominated in the supply of sediment to the broader and more even marshes.

Modern and contemporary marshes

The narrow strips of unreclaimed wetland have acted through modern times up to the present as sites of active mudflats and marsh sedimentation, under the control of processes operating on a decadal to century time scale. Allen and Rae (1987) recognised that the active wetland was underlain by three erosively based, offlapping morphostratigraphic units (Fig. 7.2) – the Rumney, Awre, and Northwick Formations – the upper surfaces of which descended stairlike toward the sea (Figs 7.6–7.8). These system-wide units of estuarine silt began to form at different times but, because each lay within the range of the tide, they simultaneously continued to receive sediment, albeit at a rate varying inversely as surface elevation in the tidal frame.

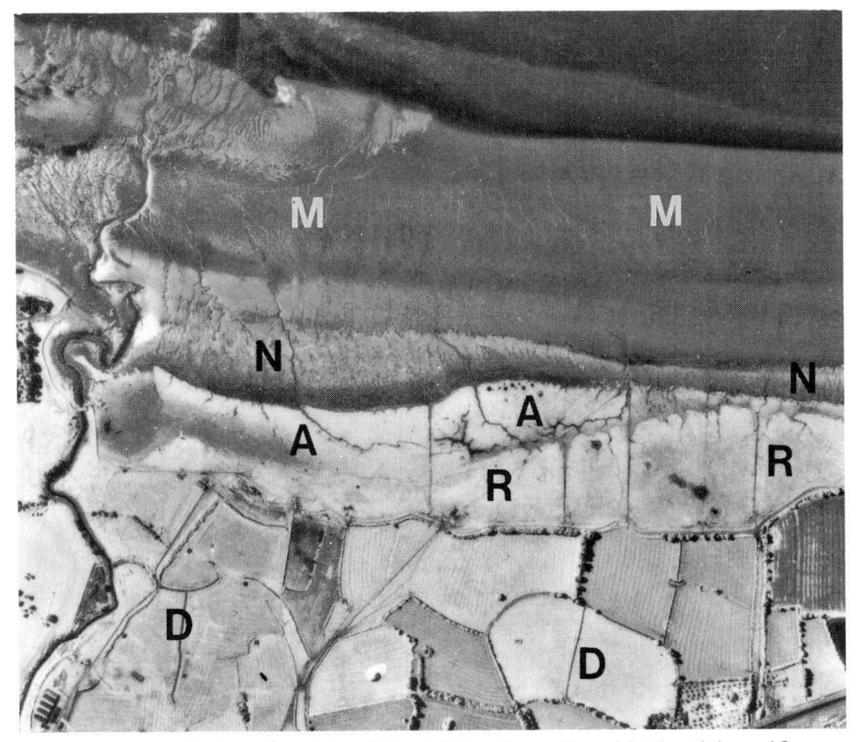

Fig. 7.6. Aerial view of reclaimed land (D), saltmarsh and high tidal mudflats (M) at Northwick Warth, Avon. The terraces on the marsh are underlain by the Rumney (R), Awre (A) and Northwick (N) Formations. Area measures approximately 1400 × 1250 m. Crown copyright reserved.

The active marshes are simple morphologically and botanically. Typically, they lack networks of creeks and gullies (e.g. Fig. 7.6), although very locally (e.g. Rumney Great Wharf) a few generally unconnected meandering gullies can be seen, together with scattered salt pans. At an average spacing along the coastline of a few kilometres, the marshes are crossed by deep tidal channels (pills) which are all that remain after reclamation and sluice-building of the natural drainage lines which emptied fresh water across and from the wetlands (Allen, 1985b). The mud settling on the sides of the pills at high tide is carried downslope at low water by a variety of mass-movement processes and eventually removed by the small freshwater discharges (Fig. 7.9), which thus combine to keep the features open. Some of the mass movements seen along the pills involve substantial volumes of relatively consolidated estuarine sediment. These may in part record a tendency for the pills to widen as relative sea level rises

Fig. 7.7. Marsh terraces underlying the Rumney (R), Awre (A) and Northwick (N) Formations, Northwick Warth, Avon. Spade for scale.

Fig. 7.8. Erosional contact in shore-normal section exposing the Northwick (above base of spade) and Rumney (below) Formations, Chittening Warth, Avon. Spade for scale.

Fig. 7.9. Right bank of Littleton Pill, Avon, showing evidence of small-scale (claw-like features) and large-scale (curved head fractures) mass movements. Creek about 8 m deep below marsh.

in the area. The marshes, which have long been grazed, belong floristically to Chapman's (1960) western group. Broadly, three botanical zones – *Festuca, Puccinellia* and *Salicornia/Zostera/Spartina* – can be found in descending order (Smith, 1979). *Spartina* has become increasingly dominant since its introduction almost 80 years ago, and is now in many places the characteristic species of the lower marshes. Its fortunes, however, have recently waned considerably.

Except at the very few surviving places where the accumulation of the Wentlooge Formation appears to have continued to the present (e.g. Tidenham, Woolaston), the highest marsh is underlain by the Rumney Formation (Allen and Rae, 1987; Allen, 1987b), composed of pale brown grading up into mid and then dark grey silts. The Rumney Formation appears to have begun to form no earlier than the late seventeenth or early eighteenth century, to judge from the evidence of ceramics preserved at the erosional base (e.g. Allen and Fulford, 1988), a scattering of early maps, and the restriction of anthropogenically polluted sediments to a comparatively thin uppermost interval (Allen and Rae, 1986, 1987; Allen, 1987c, 1988). The

Awre Formation, underlying the middle marsh, is a sequence of mid to dark grey silts which on documentary and geochemical grounds began to accumulate in the late nineteenth century. Substantial anthropogenic pollution of the system had already occured by the start of depositon, but the formation includes high up the reversal of geochemical trends which marks the boundary between Allen and Rae's (1986, 1987) Chemozones II and III, dated to *c.* 1945. The *Spartina*-dominated low marsh is underlain by the Northwick Formation, a second sequence of chiefly dark grey silts. Geochemical, documentary and ceramic evidence assigns the start of deposition to the second quarter of the present century, coincidentally with the main introduction of *Spartina* into the area. The cliff which forms the inland margin of the Northwick Formation is known from documentary evidence to have retreated inland at a rate of the order of 1 m a^{-1}.

Although not yet conclusively identified, the factors responsible for these morphostratigraphic units must be capable of explaining (1) the system-wide distribution of the deposits (units are missing here and there because of local influences, such as channel wandering), and (2) the alternation of platform- and cliff-cutting with mudflat-marsh deposition. There is no evidence that sea level in the area varied in a stepped fashion (Woodworth, 1987, 1990) and, although marshes tend to become cliffed as they ascend the tidal frame (Yapp *et al.*, 1917), the growth of relief provides no mechanism for the return of accretion. Considering that the tidal waters hold the equivalent of some 10–20 years of fluvial supply of fine sediment (Collins, 1983), it is difficult to see how fluctuations in the discharge of suspended load by the rivers can explain the offlapping morphostratigraphic units. The most plausible interpretation of the sequence seems to be that it records changes in regional wind-wave climate, an increase in severity promoting the direct quarrying of mud cliffs and shoreline platforms and the resuspension of mud settled out at high tide, but a decline favouring a regime of net accretion (Allen and Rae, 1987; Allen, 1987b). The instrumental record of winds in the British Isles is comparatively short and difficult to interpret, but it is perhaps no coincidence that there has been a substantial statistical decline in the annual frequency of days of westerly and southwesterly winds over the time span represented by the Northwick Formation (Lamb, 1982). In the same vein, the Rumney Formation is readily seen as reflecting the decay of the cool and apparently stormy conditions of the Little Ice Age (Grove, 1988).

It is at present unclear whether or not the Wentlooge Formation

Fig. 7.10. Preserved seasonal banding in Awre Formation, Tites Point, Gloucestershire. Spade for scale.

includes stratigraphic elements originating in the same way as these morphostratigraphic units of the last few hundred years. It may be possible to identify them, however, by the presence of a characteristic, probably seasonal banding which declines in thickness upward above an erosional contact. As seen in the Rumney, Awre and Northwick Formations (Fig. 7.10), the banding consists of an alternation on a decimetre to centimetre scale of bundles of relatively coarse-grained laminae (?winter-spring) with groups of comparatively fine-grained ones (?summer-autumn). On a finer scale in the field, the banding is difficult to distinguish from among the general lamination of the marsh deposits.

A model for mudflat-marsh vertical growth

The presence of the Rumney, Awre and Northwick Formations throughout the inner Bristol Channel and Severn Estuary, whatever their true explanation, points to the great sensitivity of the high-tide shoreline to small departures from the long-term norm amongst the controlling factors. An insight into at least the accretionary response of the system can be gained from a simple, one-dimensional model for the vertical growth of a mudflat-marsh for which a partial validation has been provided (Allen, 1990d, 1990e, 1990f). Because of the ways in which sediment is supplied to mudflats and tidal marshes, the natural frame of reference in which to consider upward growth is a tidal one, with datum at the level of the lowest astronomical tide (or tide-gauge datum, its practical equivalent). This datum moves compared to the Earth as relative mean sea level and tidal regime change. The local growth equation has the form

$$\frac{\mathrm{d}E}{\mathrm{d}t} = \frac{\mathrm{d}S_{\mathrm{min}}}{\mathrm{d}t} + \frac{\mathrm{d}S_{\mathrm{org}}}{\mathrm{d}t} + \frac{\mathrm{d}A}{\mathrm{d}t} - \frac{\mathrm{d}M}{\mathrm{d}t} - \frac{\mathrm{d}P}{\mathrm{d}t} \tag{1}$$

in which dE is the change in the elevation E of the sediment surface relative to tidal datum, t is time, $\mathrm{d}S_{\mathrm{min}}$ the thickness of minerogenic sediment added by the tide, $\mathrm{d}S_{\mathrm{org}}$ the thickness of added organic sedimentary material, dA the change in amplitude of the extreme astronomical tide (increase positive), dM the change in relative sea level (upward positive), and dP the change in the position of the sediment surface due to load-related consolidation of the above-bedrock estuarine sequence. The first three terms refer in different ways to the tidal regime, the fourth to an external forcing factor, and the fifth (bound up with the fourth) to an initial condition.

The minerogenic term may be modelled by noting again the morphological and hydraulic analogy between tidal mudflats and marshes and river flood plains, to which fine sediment is supplied by diffusion and convection, but with the additional factor that the invading waters become slack for a limited period around to soon after high tide. Hence the added thickness of minerogenic sediment should increase with the extent of tidal wetting and with the concentration and settling velocity characteristic of the suspended mud. For any year we may write, neglecting storm effects,

$$\frac{\mathrm{d}S_{\mathrm{min}}}{\mathrm{d}t} = \mathrm{k}\,\frac{1}{1-p}\sum^{\mathrm{year}}\sum^{T} \mathrm{C}(H,t_{\mathrm{r}}).\,W(t_{\mathrm{r}}) \tag{2}$$

where $0 \leqslant k \leqslant 1$ is the fraction of the sediment thickness remaining at the close of the year after seasonal and shorter-term compaction due to drying, p the fractional porosity of this deposit, T the duration of wetting of the sediment surface by a single astronomical tide (implicitly determined by tidal height and the elevation E), C a characteristic fractional volume concentration of mineral matter in the tidal waters, H the height of the tide, t_r the sequential position of the tide in the chosen year, and W a characteristic terminal settling velocity for the suspended mineral particles. The inner summation sign implies the evaluation of $C\ (H,\ t_r).\ W(t_r)$ over a single tide, and the outer one the accumulation of the effects of all of the tides of the year. As a representative calculation shows (Allen, 1990d), the minerogenic term is a steeply decreasing, non-linear function of surface elevation and of mudflat-marsh age (Fig. 7.11a).

The organogenic term depends on the amount of root matter and surface litter accumulated from indigenous marsh plants. As these are strongly zoned intertidally (Smith, 1979; Long and Mason, 1983), the organogenic term is probably an increasing but weak function of marsh elevation. It appears, from the thicknesses and age ranges of the mid-Flandrian peats (e.g. Heyworth and Kidson, 1982; Smith and Morgan, 1989), that the order of magnitude of the organogenic term in the Severn Estuary and inner Bristol Channel is unlikely to exceed 1 mm a^{-1}.

The amplitude term is included in Eq. (1) because, in an area like the inner Bristol Channel and Severn Estuary, effectively at the head of a lengthy gulf on the inner margin of a broad but complex continental shelf, the tidal regime is likely to be affected by changes in relative sea level. Tooley (1985) reports calculations suggesting that the tidal range in the inner Bristol Channel is increasing very gradually as the sea deepens. Woodworth *et al.* (1991) find evidence from tide-gauge records for a slight increase in range over recent decades. Measurements of marsh accretion in the area (Allen and Rae, 1988; Allen, 1991) can be suggested to mean that the effect has become important in recent centuries, although other interpretations of the data are possible.

The sea-level term currently takes substantial but variable values over southern Britain. In the southwest, relative mean sea level appears currently to be rising at a rate approaching at least 2 mm a^{-1} (Woodworth, 1987; Shennan, 1989a).

The consolidation term is formally necessary but in practice it may be possible to neglect it. Drying during tidal exposure, and seasonal drying related to associations of high temperature and little rainfall with weak spring tides, combine to ensure that the deposits formed on high tidal mudflats and tidal marshes along the inner Bristol Channel and Severn Estuary are already quite well consolidated

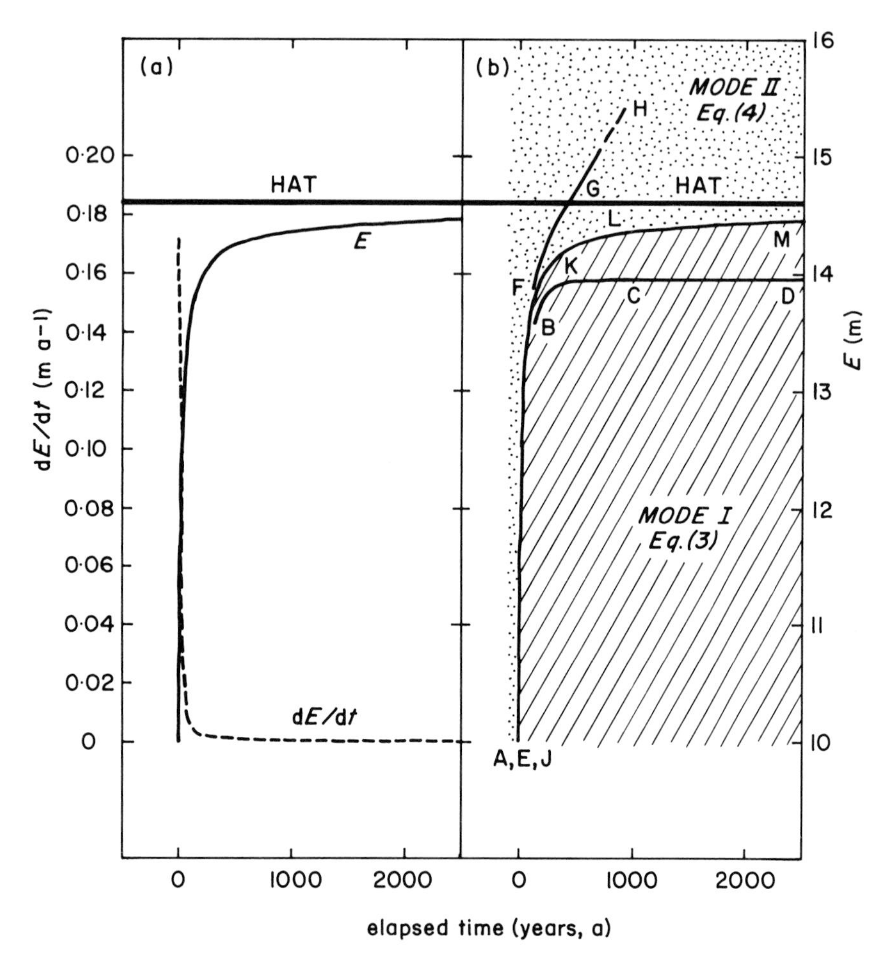

Fig. 7.11. Representative calculations for saltmarsh growth in the Severn Estuary. Curve ABCD: $dS_{org}/dt=0$, $dA/dt=0$, $dM/dt=2\times10^{-3}$ m a^{-1}, $dP/dt=0$. Curve EFGH: $dSorg/dt=2\times10^{-3}$ m a^{-1}, $dA/dt=0$, $dM/dt=0$, $dA/dt=0$. Curve JKLM: all terms except $dS_{min/dt}$ set at zero. Origin is tide-gauge datum 0.2 m above lowest astronomical tide.

before acquiring a significant overburden. Hawkins *et al.* (1989) found the post-glacial estuarine silts of the area to be well consolidated and, at certain levels, over-consolidated.

The above deceptively simple model has several implications for marsh growth, type (minerogenic or organogenic), and attainment of, and elevation at, dynamic equilibrium ($dE/dt=0$) (Allen, 1990d, 1990e, 1990f). Mode I growth (curve ABCD in Fig. 7.11b), favoured by a rising sea level, occurs where

$$\left(\frac{dS_{org}}{dt} + \frac{dA}{dt} - \frac{dM}{dt} - \frac{dP}{dt} \right) < 0 \quad (3)$$

and creates a marsh in dynamic equilibrium at a surface elevation *less* than that of the highest astronomical tide after a maturation time inversely related to k, C and W in Eq. (2). Such marshes will not be as effective in absorbing wave energy during coastal defence as might otherwise be the case. The elevation assumed by the equilibrium marsh, equalling that at which the minerogenic term just balances the remaining terms to the right in Eq. (1), declines relative to the position of the highest astronomical tide as k, C and W fall in value. Representative calculations for the inner Bristol Channel and Severn Estuary point to maturation times of the order of 10^2–10^3 years and to equilibrium surface elevations approaching one metre below the level of the highest astronomical tide. The mud-dominated character of the Wentlooge Formation, and the minerogenic nature of the marshes formed over the last few centuries, is fully in accord with Eq. (3), given the long-term trend in sea-level (Fig. 7.2d). Mode II growth (curve EFGH in Fig. 7.11b) occurs where

$$\left(\frac{dS_{org}}{dt} + \frac{dA}{dt} - \frac{dM}{dt} - \frac{dP}{dt} \right) > 0 \quad (4)$$

dynamic equilibrium being unattainable, since the minerogenic term cannot be negative. The curve JKLM (Fig. 7.11b), given by the condition

$$\left(\frac{dS_{org}}{dt} + \frac{dA}{dt} - \frac{dM}{dt} - \frac{dP}{dt} \right) = 0 \quad (5)$$

defines the boundary between mode I and mode II growth; dynamic equilibrium is possible, but only after an infinitely long growth period. Mode II, favoured by a falling or near-stable sea level, is clearly one eventually permitting the formation of an organogenic (i.e. peat-forming) marsh. Peats can also arise under mode I, but only where the minerogenic term, as limited by the sum of the amplitude, sea-level and consolidation terms, is sufficiently small in value. The implied surface elevations of such organogenic equilibrium marshes are comparatively very high in the tidal frame, and lend no credence to the choice of indicative meanings traditionally given on ecological grounds to radiocarbon-dated peats in sea-level studies (e.g. Shennan,

1986a, 1986b). A further implication of the model is that the response of a mudflat or marsh will lag behind any unsteady change of relative sea level, an inference already with some empirical support (Shennan, 1989b). The model thus suggests that the interleaved peats and silts of the middle Wentlooge Formation record, in some cases, variations on the theme of mode I growth and, in others, an alternation of mode I and mode II growth, all in response probably to an unstable sea level.

Conclusion

The deposits of high tidal mudflats and marshes have accumulated to a considerable extent over post-glacial time in the inner Bristol Channel and Severn Estuary. Although there has been much reclamation in the area, beginning in, and more than half completed by, the Roman period, mudflats and marshes continue to occupy an important place in the system today. The marshes, apparently simple morphologically in comparison with many others in Britain, have varied in time and space between minerogenic (predominantly) and organogenic. Except during a lengthy mid-Flandrian epoch, when the upward movement of relative sea level experienced an unusual instability, the wetlands were exclusively minerogenic, and this character marks the present marsh environment. A partly validated one-dimensional model for mudflat-marsh growth in the area indicates that the modern marshes are grading to an elevation in the tidal frame which is substantially less than that of the highest astronomical tide, and that this shortfall will worsen should the rate of sea-level rise increase above the current value. These marshes will be less effective in coastal defence than they might otherwise have been. Another factor limiting the usefulness of the mudflats and marshes in defence is their sensitivity on a decadal-century timescale to an erosion-accretion cycle apparently governed by changes in wind-wave climate.

References

ALLEN, J.R.L. (1985a) *Principles of Physical Sedimentology.* Allen & Unwin, London.

ALLEN, J.R.L. (1985b) Intertidal drainage and mass-movement processes in the Severn Estuary: rills and creeks (pills). *Journal of the Geological Society, London* **142**, 849–861.

ALLEN, J.R.L. (1986) A short history of saltmarsh reclamation at Slimbridge Warth and neighbouring areas, Gloucestershire. *Transactions of the Bristol and Gloucestershire Archaeological Society* **104**, 139–155.

ALLEN, J.R.L. (1987a) Dimlington Stadial (late Devensian) ice-wedge casts and involutions in the Severn Estuary, southwest Britain. *Geological Journal* **22**, 109–118.

ALLEN, J.R.L. (1987b) Late Flandrian shoreline oscillations in the Severn Estuary: the Rumney Formation at its typesite (Cardiff area). *Philosophical Transactions of the Royal Society* **B315**, 157–184.

ALLEN, J.R.L. (1987c) Towards a quantitative chemostratigraphic model for the sediments of late Flandrian age in the Severn Estuary, U.K. *Sedimentary Geology* **53**, 73–100.

ALLEN, J.R.L. (1988) Modern-period muddy sediments in the Severn Estuary (southwestern UK): a pollutant-based model for dating and correlation. *Sedimentary Geology* **58**, 1–21.

ALLEN, J.R.L. (1990a) The Severn Estuary in southwest Britain: its retreat under marine transgression, and fine-sediment regime. *Sedimentary Geology* **96**, 13–28.

ALLEN, J.R.L. (1990b) Reclamation and sea defence in Rumney Parish (Monmouthshire). *Archaeologia Cambrensis* **97**, 135–140.

ALLEN J.R.L. (1990c) Late Flandrian shoreline oscillations in the Severn Estuary: change and reclamation at Arlingham, Gloucesterhire. *Philosophical Transactions of the Royal Society* **A330**, 315–334.

ALLEN, J.R.L. (1990d) Saltmarsh growth and stratification: a numerical model with special reference to the Severn Estuary, southwest Britain. *Marine Geology* **95**, 77–96.

ALLEN, J.R.L. (1990e) Constraints on measurement of sea-level movements from saltmarsh accrection rates. *Journal of the Geological Society, London* **147**, 5–7.

ALLEN, J.R.L. (1990f) The formation of coastal peat marshes under an upward tendency of relative sea-level. Journal of the Geological Society, London **147**, 743–745.

ALLEN, J.R.L. (1991) Saltmarsh accretion and sea-level movement in the inner Severn Estuary: the archaeological and historical contribution. *Journal of the Geological Society, London* **148**, 485–494.

ALLEN, J.R.L. (1992) Fine sediment and its sources, Severn Estuary and inner Bristol Channel, southwest Britain. *Sedimentary Geology* (in press).

ALLEN, J.R.L. & FULFORD, M.G. (1986) The Wentlooge Level: a Romano-British saltmarsh reclamation in southeast Wales. *Britannia* **17**, 91–117.

ALLEN, J.R.L. & FULFORD, M.G. (1988) Romano–British settlement and industry on the wetlands of the Severn Estuary. *Antiquaries Journal* **62**, 237–289.

ALLEN, J.R.L. & FULFORD, M.G. (1990) Romano–British and later reclamations on the Severn salt marshes in the Elmore area, Gloucestershire. *Transactions of the Bristol and Gloucestershire Archaeological Society* (in press).

ALLEN, J.R.L. & FULFORD, M.G. (1991) Romano–British reclamations at Longney, Gloucestershire, and the evidence for early settlement of the inner Severn Estuary. *Antiquaries Journal* (in press).

ALLEN, J.R.L. & RAE J.E. (1986) Time sequence of metal pollution, Severn Estuary, southwestern UK. *Marine Pollution Bulletin* **17**, 427–431.

ALLEN, J.R.L. & RAE, J.E. (1987) Late Flandrian shoreline oscillations in the Severn Estuary: a geomorphological and stratigraphical reconnaissance. *Philosophical Transactions of the Royal Society* **B315**, 185–230.

ALLEN, J.R.L. & RAE, J.E. (1988) Vertical saltmarsh accretion since the Roman period in the Severn Estuary, southwest Britain. *Marine Geology* **83**, 225–235.

ANDERSON, J.G.C. (1968) The concealed rock surface and overlying deposits of Severn Valley from Upton to Neath. *Proceedings of the South Wales Institution of Engineers* **83**, 27–47.

ANDERSON, J.G.C. (1974) The buried channels, rock floors and rock basins, and overlying deposits of the South Wales valleys from Wye to Neath. *Proceedings of the South Wales Institution of Engineers* **88**, 11–25.

ANDERSON, J.G.C. & BLUNDELL, C.R.K. (1965) The sub-drift rock-surface and buried valleys of the Cardiff district. *Proceedings of the Geologists' Association* **76**, 367–378.

ANDREWS, J.T., GILBERTSON, D.D. & HAWKINS, A.B. (1984) The Pleistocene succession of the Severn Estuary: a revised model based upon amino acid racemization studies. *Journal of the Geology Society, London* **141**, 967–974.

ASTILL, G. (1988) Fields. In G. Astill & A. Grant (eds) *The Countryside of Medieval England.* Blackwell, Oxford, 62–85.

CHAPMAN, V.J. (1960) *Salt Marshes and Salt Deserts of the World.* Leonard Hill, London, 392 pp.

CODRINGTON, T. (1898) On some submerged rock-valleys in South Wales, Devon and Cornwall. *Quarterly Journal of the Geological Society of London* **54**, 251–276.

COLLINS, M.B. (1983) Supply, distribution, and transport of suspended sediment in a macrotidal environment: Bristol Channel, U.K. *Canadian Journal of Fisheries and Aquatic Sciences* **40** (Supplement No. 1), 44–59.

COOK, N.J. & PRIOR, M.J. (1987) Extreme wind climate of the United Kingdom. *Journal of Wind Energy and Industrial Aerodynamics* **26**, 371–389.

COLES, B. & COLES, J.M. (1986) *Sweet Track to Glastonbury: the Somerset Levels in Prehistory.* Thames and Hudson, London, 200 pp.

CRICKMORE, M.J. (1982) Data collection, – tides, tidal currents, suspended sediment. In Institution of Civil Engineers (eds) *Severn Barrage.* Thomas Telford, London, 19–26.

CUNLIFFE, B.W. (1966) The Somerset Levels in the Roman period. *Council for British Archaeology Research Report* **7**, 68–73.

FUNNELL, B.M. & PEARSON, I. (1989) Holocene sedimentation on the north Norfolk barrier coast in relation to relative sea-level change. *Journal of Quaternary Science* **4**, 25–36.

GRAY, A.J. (1979) The ecology of Morecambe Bay. V. The salt marshes of Morecambe Bay. *Journal of Applied Ecology* **9**, 207–220.

GRAY, A.J. & SCOTT, R. (1987) Salt marshes. In N.A. Robinson & A.W. Pringle (eds) *Morecambe Bay: an Assessment of present Ecological Knowledge.* Morecambe Bay Study Group and Centre for North West Regional Studies, Lancaster, 97–117.

GREEN, H.S. (1989) Some recent archaeological and faunal discoveries from the Severn Estuary levels. *Bulletin of the Board of Celtic Studies* **36**, 187–199.

GROVE, J.M. (1988) *The Little Ice Age.* Routledge and Kegan Paul, London, 500 pp.

HALL, D. (1981) The origins of open-field agriculture – the archaeological fieldwork evidence. In T. Rowley (ed.) *The Origins of Open-Field Agriculture.* Croom Helm, London, 22–38.

HALL, D. (1982) *Medieval Fields.* Shire Publications, Princes Risborough, 56 pp.

HAMILTON, D. (1979) The high energy, sand and mud regimes of the Severn Estuary, S.W. Britain. In R.T. Severn, D.L. Dineley & L.E. Hawker (eds) *Tidal Power and Estuary Management.* Scientechnica, Bristol, 162–172.

HAWKINS, A.B. (1962) The buried channel of the Bristol Avon. *Geological Magazine* **99**, 369–374.

HAWKINS, A.B., LARNACH, W.J., LLOYD, I.M. & NASH, D.F.T. (1989) Selecting the location, and the initial investigation of the SERC soft clay test site. *Quarterly Journal of Engineering Geology* **22**, 281–316.

HEYWORTH, A. & KIDSON, C. (1982) Sea-level changes in southwest England and Wales. *Proceedings of the Geologists' Association* **93**, 91–111.

HYDROGRAPHER OF THE NAVY (1990) *Admiralty Tide Tables. Vol. 1. 1991. European Waters including the Mediterranean Sea.* Hydrographer of the Navy, Taunton, 438 pp.

JEFFERIES, R.L., WILLIS, A.J. & YEMM, E.W. (1968) The late- and post-glacial history of the Gordano Valley, northwest Somerset. *New Phytologist* **67**, 335–348.

KNIGHT, D.W. (1989) Hydraulics of flood channels. In K. Bevin & P. Carling (eds) *Floods: Hydrological, Sedimentological and Geomorphological Implications.* Wiley, Chichester, 83–105.

LAMB, H.H. (1982) *Climate, History and the Modern World.* Methuen, London, 387 pp.

LENNON, G.W. (1963a) A frequency investigation of abnormally high tidal levels at certain west coast ports. *Proceedings of the Institution of Civil Engineers* **25**, 451–484.

LENNON, G.W. (1963b) The identification of weather conditions associated with the generation of major storm surges along the west coast of the British Isles. *Quarterly Journal of the Royal Meteorological Society* **89**, 381–394.

LOCKE, S. (1971) The post glacial deposits of the Caldicot Level and associated archaeological discoveries. *Monmouthshire Antiquary* **3**, 1–16.

LONG, S.P. & MASON, C.F. (1983) *Saltmarsh Ecology*. Blackie, London, 160 pp.

MARSHALL, D.R. (1962) The morphology of the upper Solway salt marshes. *Scottish Geographical Magazine* **78**, 81–99.

MCDONNELL, R.R.J. (1985) *Archaeological Survey of the Somerset Claylands, Part I*. Somerset County Council, Taunton, 13 pp.

MCDONNELL, R.R.J. (1986) *Archaeological Survey of the Somerset Claylands, Part II*. Somerset County Council, Taunton, 11 pp.

PIZZUTO, J.E. (1987) Sediment diffusion during overbank flows. *Sedimentology* **34**, 301–317.

SEDDON, B. (1964) Submerged peat layers in the Severn channel near Avonmouth. *Proceedings of the Bristol Naturalists' Society* **31**, 101–106.

SHENNAN, I. (1983) Flandrian and late Devensian sea-level changes and associated coastal movements in England and Wales. In D.E. Smith & A.G. Dawson (eds) *Shorelines and Isostasy*. Academic Press, London, 255–283.

SHENNAN, I. (1986a) Flandrian sea-level changes in the Fenland. I: the geological setting and evidence for relative sea-level changes. *Journal of Quaternary Science* **1**, 119–154.

SHENNAN, I. (1986b) Flandrian sea-level changes in the Fenland. II: tendencies of sea-level movement, altitudinal changes, and local and regional factors. *Journal of Quaternary Science* **1**, 155–179.

SHENNAN, I. (1989a) Holocene crustal movements and sea-level changes in Great Britain. *Journal of Quaternary Science* **4**, 77–89.

SHENNAN, I. (1989b) Holocene sea-level changes and crustal movements in the North Sea region: an experiment with regional eustasy. In D.B. Scott, P.A. Pirazzoli & C.A. Honig (eds) *Late Quaternary Sea Level: Correlation and Applications*. Kluwer, Dordrecht, 1–25.

SHUTTLER, R.M. (1982) The wave climate in the Severn Estuary. In Institution of Civil Engineers (eds) *Severn Barrage*. Thomas Telford, London, 27–34.

SMITH, A.G. & MORGAN, L.A. (1989) A succession to ombrotrophic bog in the Gwent Levels, and its demise: a Welsh parallel to the peats of the Somerset Levels. *New Phytologist* **112**, 145–167.

SMITH, L. (1979) *A Survey of Salt Marshes in the Severn Estuary*. Nature Conservancy Council, London, 100 pp.

SMITH, S.G. (1983) The seasonal variation of wind speed in the United Kingdom. *Weather* **38**, 98–103.

STEERS, J.A. (1960) *Scolt Head Island.* Heffer, Cambridge, 269 pp.

STEPHENS, C.V. (1986) A three-dimensional model for tides and salinity in the Bristol Channel. *Continental Shelf Research* **6**, 531–560.

TOOLEY, M.J. (1985) Sea levels. *Progress in Physical Geography* **9**, 113–120.

UNCLES, R.J. (1982) Computed and observed residual currents in the Bristol Channel. *Oceanologia Acta* **5**, 11–20.

UNCLES, R.J. (1984) Hydrodynamics of the Bristol Channel. *Marine Pollution Bulletin* **15**, 47–53.

UNCLES, R.J., JORDAN, M.B. & TAYLOR, A.H. (1986) Temporal variability of elevations, currents and salinity in a well-mixed estuary. In D.A. Wolfe (ed.) *Estuarine Variability.* Academic Press, Orlando, 103–122.

WHITTAKER, A. & GREEN, G.W. (1983) The geology of the country around Weston-super-mare. *Memoirs of the Geological Survey of Great Britain.*

WHITTLE, A., ANTOINE, S., GARDINER, N., MILLES, A. & WEBSTER, P. (1989) Two late Bronze Age occupations and an Iron Age channel on the Gwent foreshore. *Bulletin of the Board of Celtic Studies* **36**, 200–223.

WILLIAMS, D.J. (1968) The buried channel and superficial deposits of the lower Usk and their correlation with similar features in the lower Severn. *Proceedings of the Geologists' Association* **79**, 325–348.

WILLIAMS, M. (1970) *The Draining of the Somerset Levels.* Cambridge University Press, Cambridge, 288 pp.

WILLS, L.J. (1938) The Pleistocene development of the Severn from Bridgnorth to the Sea. *Quarterly Journal of the Geological Society of London* **94**, 161–242.

WOODWORTH, P.L. (1987) Trends in U.K. mean sea level. *Marine Geodesy* **11**, 57–87.

WOODWORTH, P.L. (1990) A search for accelerations in records of European mean sea-level. *International Journal of Climatology* **10**, 129–143.

WOODWORTH, P.L., SHAW, S.M. & BLACKMAN, D.L. (1991) Secular trends in mean tidal range around the British Isles and along the adjacent European coastline. *Geophysical Journal International* **104**, 593–609.

YAPP, R.H., JOHNS, D. & JONES, O.T. (1917) The salt marshes of the Dovey Estuary. Part II. The salt marshes. *Journal of Ecology* **5**, 65–103.

8

Saltmarshes on the barrier coastline of North Norfolk, eastern England

K. PYE

Introduction

Saltmarshes fringe much of the north Norfolk coast between Holme-next-the-Sea and Cley-next-the-Sea, a distance of some 35 km (Fig. 8.1). In places the belt of marshes is more than 2 km wide, and the total area of saltmarsh exceeds 3000 ha. In addition, there are approximately 6000 ha of mudflats and sandflats exposed at low tide, and more than 1500 ha of enclosed fresh and brackish marshes which have been reclaimed from the sea since the mid-seventeenth century. The area represents one of the finest natural coasts in Britain and is of exceptional importance to both naturalists and specialists in many branches of science. It is an important breeding and migration stop-over site for a large number of birds and marine mammals, including oystercatchers, several species of tern, duck, geese and common seals (Seago, 1989). Much of the area has been designated as a Site of Special Scientific Interest and as a Heritage Coast. The National Trust owns ten properties between Brancaster and Weybourne, amounting to 2570 ha. All except Burnham Overy are included within the Heritage Coast and are subject to a special management plan adopted by the Trust. Perhaps the best known property, Blakeney Point, has been a nature reserve since 1912. The Nature Conservancy Council administers two National Nature Reserves, at Holkham and Scolt Head Island. The Holkham NNR, created in 1967, is the largest in England and consists of 1700 ha of marshes and dunes betwen Burnham Overy and Stiffkey belonging to the Holkham Estate, together with 2200 ha of intertidal sandflats and mudflats between Burnham Overy and Blakeney leased from the Crown Estate Commissioners. Scolt Head Island NNR was first leased to the NCC in 1953 by the National Trust and the Norfolk Naturalists Trust and is now managed by a joint committe representing the three organisations. In addition, the Norfolk Ornithologists Association has three reserves in the area, at Holme Bird Observatory

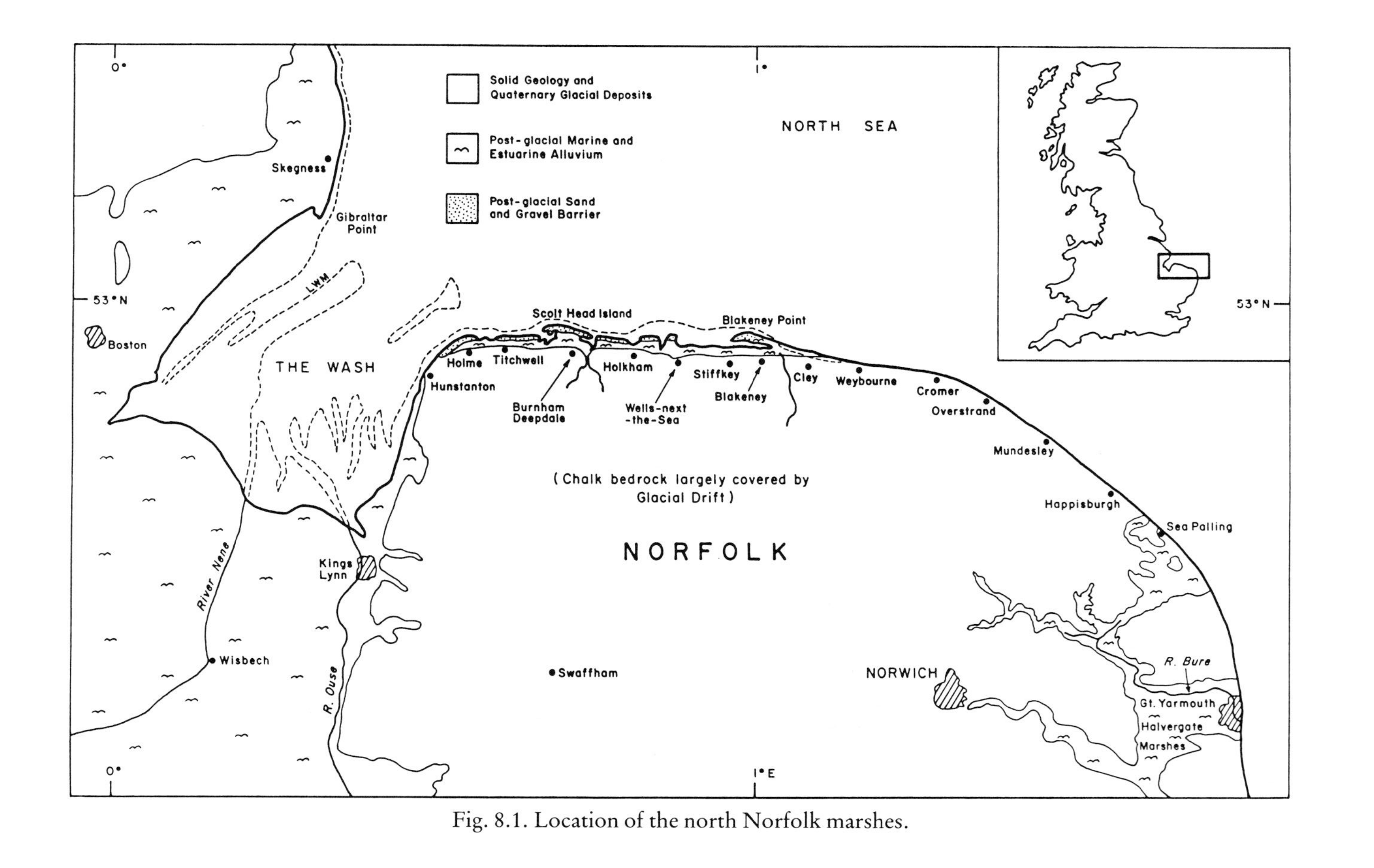

Fig. 8.1. Location of the north Norfolk marshes.

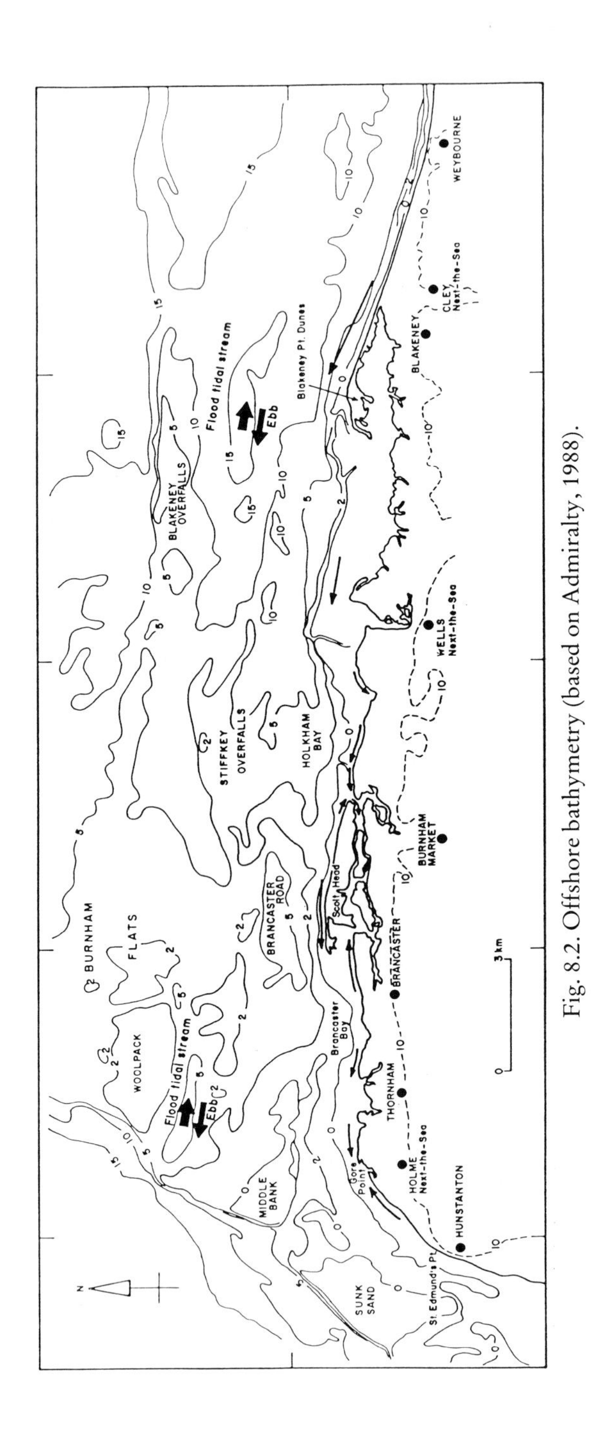

Fig. 8.2. Offshore bathymetry (based on Admiralty, 1988).

(2.5 ha), Holme Marsh Bird Reserve (36.4 ha), and Walsey Hills (1 ha), near Salthouse. The Royal Society for the Protection of Birds also administers a reserve at Titchwell Marsh (170 ha). The Norfolk Naturalists Trust owns nature reserves at Holme Dunes (175 ha) and Cley Marshes (435 ha), and adminsters an area of marsh owned by the National Trust at Gramborough Hill near Salthouse.

The north Norfolk coast has provided a training ground for generations of university students in botany, geography, zoology, geology and environmental sciences, and has been used for almost a century as a natural research laboratory by numerous distinguished scientists from the Universities of London, Cambridge and East Anglia. Amongst the most noteworthy contributions are the ecological studies by Professor F.W. Oliver and his associates, who founded the Blakeney Research Station early in this century (Oliver, 1923; Oliver and Salisbury, 1913; Carey and Oliver, 1918), the geomorphological work by Professor J.A. Steers, principally at Scolt Head Island (Steers, 1934a, 1934b, 1948, 1989), and Professor V.J. Chapman's work on saltmarsh vegetation succession and ecology (Chapman, 1938, 1959). More recent research in the area has contributed much to our understanding of the nature of tidal flows within marsh drainage systems (Bayliss-Smith *et al.*, 1979; Healey *et al.*, 1981; Green *et al.*, 1986), the factors which control spatial variations in marsh accretion rates (Pethick, 1981; Stoddart *et al.*, 1989; French, 1989; French *et al.*, 1990), geochemical changes in marsh sediments (Pye, 1981, 1984, 1988; Pye *et al.*, 1990), and many aspects of saltmarsh biology (Barnes *et al.*, 1976; Jefferies, 1977; Jefferies and Perkins 1977; Frid, 1988; Treherne and Foster, 1979; Foster, 1989; White, 1989; Woodell, 1989; Seago, 1989).

The physical environment of the North Norfolk coast

The tidal regime of The Wash is macrotidal, with a mean spring tidal range of some 6.6 m at Hunstanton (Hydrographer of the Navy, 1991), but the open water tidal range decreases quite sharply eastwards towards Cromer, where it is only 4.7 m at mean springs. The height of predicted MHWS relative to Ordnance Datum is 3.75 m at Hunstanton, 2.9 m at Wells Bar, 2.65 m at Blakeney Bar and 2.45 m at Cromer. The offshore zone between Holme and Scolt Head Island is very shallow, but to the east of Blakeney relatively deep water extends close inshore (Admiralty, 1988; Fig. 8.2). The tides inshore are subject to pronounced shallow water effects and there are marked variations in the absolute levels of high water along the coast. Within the intertidal flat and marsh zones, high water levels vary quite markedly over short distances.

The flood tidal stream off Thornham runs in a northeast to southwest direction into the Wash, having a maximum velocity of 0.6 knots at springs. The ebb tidal streams run in an easterly direction with a maximum velocity of 1.0 knots at springs. In the Blakeney area the flood tidal stream runs from east to west with a maximum velocity of 2.2 knots at springs, while the ebb tidal stream runs in the opposite direction with a peak spring velocity of 2.4 knots.

Wave energy along much of the coast is low to moderate, being highest east of Blakeney where deep water comes closer inshore and where there is greater exposure to winds from the northeast (direction of greatest fetch). Between Scolt and Blakeney mean wave height is 0.2–0.3 m, between Blakeney and Sheringham 0.3–0.4 m and east of Sheringham >0.4 m (Anglian Water, 1988a, 1988b). The predicted 1 in 100 year wave height between Gore Point and Sheringham is 6.0–8.0 m and to the east of Sheringham >8.0 m. Calculations indicate that the dominant waves are from the northeast and that dominant longshore wave energy from Sheringham westwards is directed towards the west (Steers, 1927; Hardy, 1964; Vincent, 1979; Anglian Water, 1988a, 1988b). Between Sheringham and Cromer the direction is variable, but to the east of Cromer it is predominantly eastwards. Vincent (1979) calculated the westward potential sand transport rate past Blakeney to be 3.5×10^5 m^3 per annum, although the actual transport rate is much less due to a deficiency of available sand-sized sediment. Grain-size trends in intertidal sands between Cromer and Hunstanton show no conclusive evidence of selective transport towards the west (McCave, 1978), possibly indicating that much of the sand is supplied from the north or west. Nearshore and shoreline features such as bars and spits indicate dominant eastward transport along much of the coast between Holme and Brancaster (Steers, 1929; Roy, 1967).

At the present day relatively little sediment is supplied to the coastal zone from east coast rivers. McCave (1987) concluded that erosion of cliffs in Norfolk and Suffolk yields an average of 7.8×10^5 t a^{-1} of fine-grained sediments to the coastal zone, compared with about 0.5×103 t a^{-1} from East Anglian rivers and 1.0×105 t a^{-1} from rivers draining into The Wash. Erosion of cliffs in Holderness yields about 1.4×106 t a^{-1} of fines. Most of the sea bed off north

Norfolk is composed of boulder clay covered by a veneer of sand or sandy gravel; it is therefore probably unimportant as a source of mud at present (McCave, 1977, 1987). Mineralogical studies have indicated that modern coastal mud deposits in The Wash and north Norfolk are relatively homogeneous, reflecting efficient mixing of fine sediment derived from several surrounding sources (Shaw, 1973). Since mud represents about one third of the sediment present in the cliffs composed of glacial till (Cambers, 1976), the total amount of sediment supplied by cliff erosion may be up to three times higher than indicated by the figures above.

Parts of the boulder clay cliffs east of Overstrand have receded by more than 180 m in the last century, although west of Overstrand the average amount of retreat has been less than 40 m (Cambers, 1976, Clayton, 1989a). The average annual input of sediment reaching the north Norfolk coast from this source since 1885 is estimated to be about 0.6×105 m3 a^{-1}, which compares with an estimated 2.6×105 m^3 a^{-1} transported towards Great Yarmouth (Clayton, 1989a). Most of the mud supplied to The Wash and north Norfolk may therefore originate in north Lincolnshire and Holderness, being transported southwards mainly by tidal currents (Clayton, 1976). There is no difficulty concerning the immediate source of sand supplied to the north Norfolk coast since much of floor of the adjacent North Sea is covered by at least 1 m of sandy deposits.

Physiography and Holocene evolution of the coast

The north Norfolk coast provides an example of saltmarsh development on a barrier coast undergoing gradual submergence but where landward movement of the coastal sedimentary units is constrained by relatively high ground to the south. The Holocene coastal sediments have accumulated seawards of a degraded cliff-line cut into the Chalk and late Pleistocene glacial drift deposits which can be traced from near Holme-next-the Sea to Salthouse (Woodward, 1882). A gravel beach of inferred last interglacial age is preserved immediately landward of the marshes between Stiffkey and Morston (Solomon, 1931; Gale *et al.*, 1988) and near Hunstanton (Gallois, 1976). These fossil beach deposits are overlain by glacial till of last glacial (Devensian) age and themselves overlie a chalky till unit which is probably of penultimate glacial (Wolstonian) age (Gale *et al.*, 1988). The adjacent floor of the North Sea was glaciated in the last ice age,

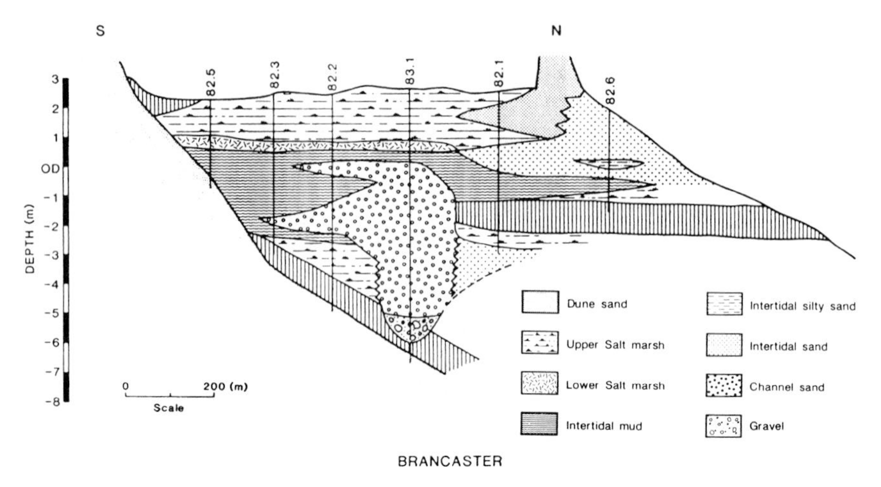

Fig. 8.3. Cross-section showing the stratigraphy of the Brancaster marshes (after Funnell and Pearson, 1989).

when the maximum ice limit lay just south of the present coastline (Straw, 1960). The late glacial and early postglacial history of the coastal zone is poorly documented, but drilling has shown that freshwater peats formed, at least locally, on top of glacial deposits between 8400 and 4000 conventional radiocarbon years BP, when sea level was much lower and the shoreline lay well to the north of its present position (Funnell and Pearson, 1984, 1989). Saltmarsh deposits extend to a depth of at least −7.0 m OD at Stiffkey, −6.5 m OD at Holkham Gap, and −5 m OD at Brancaster, Cley and Morston (Figs 8.3 and 8.4), indicating marine incursion into parts of the present coastal zone by about 6600 years ago (Funnell and Pearson, 1989). The surface of the chalk bedrock and overlying glacial deposits dips steeply northwards in these locations. In other places, however, boomer profiling and drilling has shown that the pre-Holocene basement forms an undulating platform at relatively shallow depths below present sea level, and the thickness of postglacial marine and intertidal deposits is less than 3 m (e.g. between Burnham Overy and Holkham Gap: Anglian Water 1988a, 1988b). The pre-Holocene topography has therefore probably exerted a significant influence on the development of the present coastal features, with the first marshes being formed in a series of broad embayments which may have been partially protected by inherited glacial features and low promontories of chalk bedrock.

The inner saltmarshes at Stiffkey and Morston first began to form about 6600 conventional radiocarbon years ago, since when they have

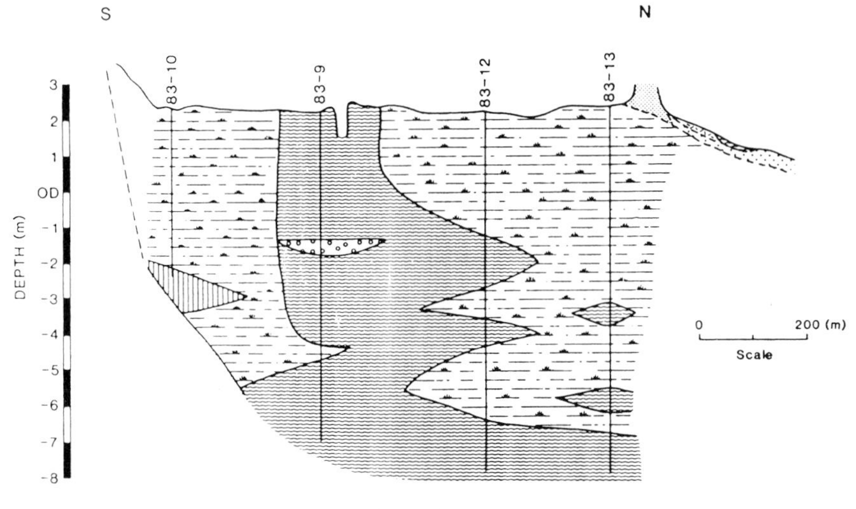

Fig. 8.4. Stratigraphy of the Upper Marsh at Stiffkey (after Funnell and Pearson, 1989).

continued to grow vertically during a period of generally rising sea level. Thin freshwater peat beds occur locally at several levels in the intertidal sediment sequence, and are exposed on the lower foreshore between Brancaster and Titchwell (Godwin, 1934, 1943; Funnell and Pearson, 1989) and on the seaward side of Scolt Head Island (Roy, 1967; Allison, 1985). However, it is uncertain whether these peat beds represent purely local variations in hydrological conditions due to geomorphological changes or short-lived falls in sea level which were regional in their effect. Funnell and Pearson (1989) recognised two main periods of peat formation, the older of which, recorded in boreholes at Brancaster and Holkham, pre-dates the postglacial marine transgression in the area. A second major episode of freshwater peat formation was identified in several places, notably at Stiffkey, Cley and Holkham, corresponding to the period 4480 ± 60 to 4880 ± 80 ^{14}C yr BP. These younger peat deposits, encountered at a depth of −3.36 to −2.93 m OD, may indicate a short-lived fall in sea level (Funnell and Pearson 1989). Another minor regression may be indicated by peats near Brancaster dated at 3470 ± 50 conventional ^{14}C yr BP (bottom of the deposit at −2.15 m OD) to 2790 ± 40 ^{14}C yr BP (top of the unit at −1.03 m OD). Since freshwater peats do not form today on this coast below an elevation of about +3.3 m OD, these peat deposits suggest an average rate of relative sea-level rise of about 1.5 mm a^{-1} during the past 2800 years. This agrees well with the regional rate of crustal subsidence in north Norfolk of 1–2 mm a^{-1} suggested by Shennan (1989).

Levelling across sea-banks at Holkham and Blakeney has also shown that marshes reclaimed in the late seventeenth century are about 600 mm lower than the actively accreting marshes on the seaward side of the embankments. Assuming that both the reclaimed and the active marshes are mature, i.e. had achieved a constant elevation relative to the moving tidal frame, this would also indicate a relative sea-level rise of about 1.5 mm a^{-1} during the past 400 years. Further evidence for a rate of sea-level rise of this magnitude is provided by sediment accretion rate measurements made on a number of marshes at Scolt Head Island since the early 1930s (Steers, 1935, 1938, 1948; Stoddart *et al.*, 1989; French, 1989). These data suggest a maximum vertical accretion rate of 5–7 mm a^{-1} on marshes which are approaching maturity. However, the sediment at the time of deposition has a low dry bulk density (0.3–0.4 t m^{-3}) and a high organic carbon content (10–20% by dry weight). On a timescale of 50–100 years the bulk density increases to 1.2–1.6 t m^{-3} due to compaction, dewatering, degradation of organic matter and dissolution of calcium carbonate (French, 1989; Pye, unpublished data). The thickness of an annual accretion lamina may therefore be reduced to about 25% of its original value over a period of about 50 years. Taking this into account, the accretion rate data from the higher marshes on Scolt Head Island are consistent with a present upward tendency of relative sea level of about 1.5 mm a^{-1}.

Tidal influence appears to have attained its maximum landward extent during the early Middle Ages, when high tides regularly flooded the Church and Slade marshes east of Wells (Purchas, 1965). The lower reaches of the Rivers Glaven, Stiffkey and Burn were also regularly inundated by high tides at this time and Holkham Lake was a tidal inlet (Hunt, 1971; Williams and Beale, 1976). Although relative sea level has apparently risen by about 750 mm in the last 500 years, tidal flooding of the coastal river valleys effectively ceased following the construction of embankments across their mouths in the seventeenth and eigtheenth centuries. Many areas of open coast marsh were also reclaimed by the construction of sea walls after about 1650, reaching a peak in the early eighteenth century. Earth embankments were constructed around Deepdale, Norton, Overy and Holkham Marshes by the Earls of Leicester and Sir Charles Turner during the period 1650–1859 (Fig. 8.5). Sea walls were also built in the mid-

Fig. 8.5. Blakeney Harbour in July 1990, viewed from the embankment which encloses the fresh marshes between Blakeney and Cley.

seventeenth century to enclose marshes between Salthouse and Blakeney (Cozens-Hardy, 1924–5), while much of the marsh between Thornham and Holme was enclosed by embankments between 1780 and 1860 (Steers, 1936). With the exception of a small area immediately to the east of Wells, none of the marshes between Wells and Blakeney has been reclaimed, although parts of Lodge Marsh were enclosed by banks and a farm established in the eighteenth century, being subsequently abandoned (Purchas, 1965; Steers, 1969).

There is considerable archival, morphological and sedimentary evidence that embanking caused enhanced siltation within the creeks in the remaining unreclaimed marshes and encouraged the accretion of younger marshes to seaward (Purchas, 1965; Steers, 1969; Cozens-Hardy, 1972; Hales and Simms, 1990). In order to improve the navigation access to the small ports of Wells, Blakeney and Cley, numerous attempts have been made to alter the drainage system of the marshes by constructing diversionary embankments, damming of certain creeks and enlarging or straightening of others.

Many of the recent saltmarshes have formed behind protective barrier islands and spits composed of gravel and sand which serve to

Fig. 8.6. The western end of Scolt Head Island, photographed from the air in June 1984, showing incipient marshes in the protected zone behind the recurved end of the island (reproduced by permission of the Cambridge University Committee for Air Photography).

reduce wave action and allow mud accumulation to take place on their landward sides (Steers, 1936, 1969; Pearson *et al.*, 1989; Fig. 8.6). Between Wells and Blakeney, however, young marshes have formed on a relatively open stretch of coast which is protected only by a wide zone of intertidal sand flats and low sand bars (Fig. 8.7). The major barrier features, including Scolt Head Island and Blakeney Spit, have clearly continued to grow in a westerly direction during the last few hundred years and have migrated landwards over the backbarrier marsh deposits (Carey and Oliver, 1918; Barfoot and Tucker, 1980; Steers, 1934a, 1989), but drilling by Allison (1985) and Funnell and Pearson (1984) suggested that in part they have been in

Fig. 8.7. The lower marsh at Warham in June 1984. The entrance to Blakeney Harbour can be seen in the top right of the photograph (reproduced by permission of the Cambridge University Committee for Air Photography).

existence for several thousands of years. The remains of an inner barrier system, which shows evidence of growth from west to east, in the marshes between Scolt and the mainland suggests that this part of the shore was protected from the dominant northeasterly waves by an outer barrier structure from an early date (Steers, 1934a, 1934b, 1969). Much of the western and central part of Scolt Head Island is underlain by gravel and sands (possibly barrier washover deposits), which in turn overlie up to 2 m of buried late Holocene marsh deposits which have been radiocarbon dated at between 6500 and 3000 conventional ^{14}C yr BP (Allison, 1985, 1989). The eastern end of the present island is probably a relatively young feature, however, formed partly by eastward sediment drift into Burnham Harbour.

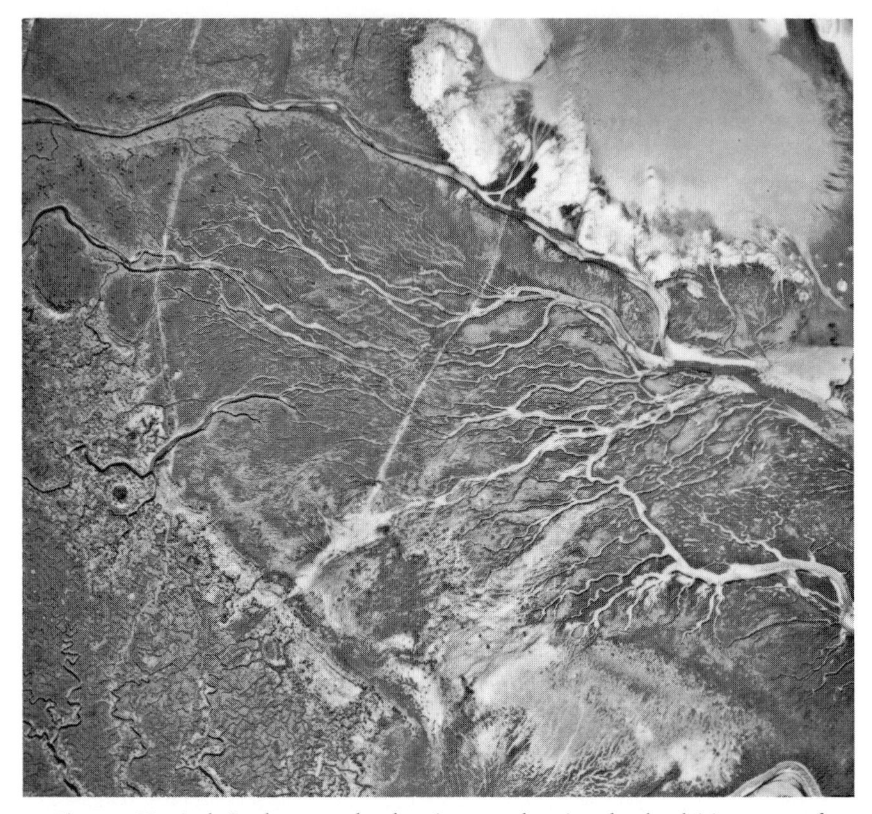

Fig. 8.8. Vertical air photograph taken in 1971 showing the dendritic nature of the creek network on Warham Lower Marsh (reproduced by permission of the Cambridge University Committee for Air Photography). Scale 1:10,000.

The latter is almost certainly a natural tidal inlet of considerable antiquity, although one which has been modified by artificial widening and deepening of the channel approaches to the harbour at Burnham Overy, and there is no strong evidence that Scolt Head Island formed from a spit which was later breached.

The marshes of north Norfolk generally are characterised by a relatively high density of drainage creeks and salt marsh pans, but there are significant differences between the drainage systems of the younger, lower marshes and the older, higher marshes (Pethick, 1969; Bayliss-Smith *et al.*, 1979). During the initial stages of marsh inception, represented by Cockle Bight Marsh on Scolt Head Island, the drainage network is poorly defined, being characterised by a broad main channel inherited from the intertidal flat stage and a relatively small number of first and second order tributaries. Large

Fig. 8.9. The Upper Marsh at Warham, looking towards Wells, taken in June 1984. Note the sinuous nature of the major drainage channel (Stonemeal Creek) and the large number of salt pans on the marsh surface (reproduced by permission of the Cambridge University Committee for Air Photography).

areas on the periphery of the youngest marshes are not drained by creeks at all. On slightly higher marshes, such as Hut Marsh on Scolt Head Island and the Lower Marsh at Warham (Fig. 8.8), the drainage system forms a better developed dendritic network. This transformation is accomplished through headward erosion and bifurcation of the first order tributaries as they extend into areas formerly drained by sheet flow (Pethick, 1969). This may represent a hydraulic readjustment to the changing flow conditions over the increasingly well vegetated marsh surface, or it may partly reflect changes in surface flow conditions associated with a transition from sand to mud deposition as the marsh grows in height.

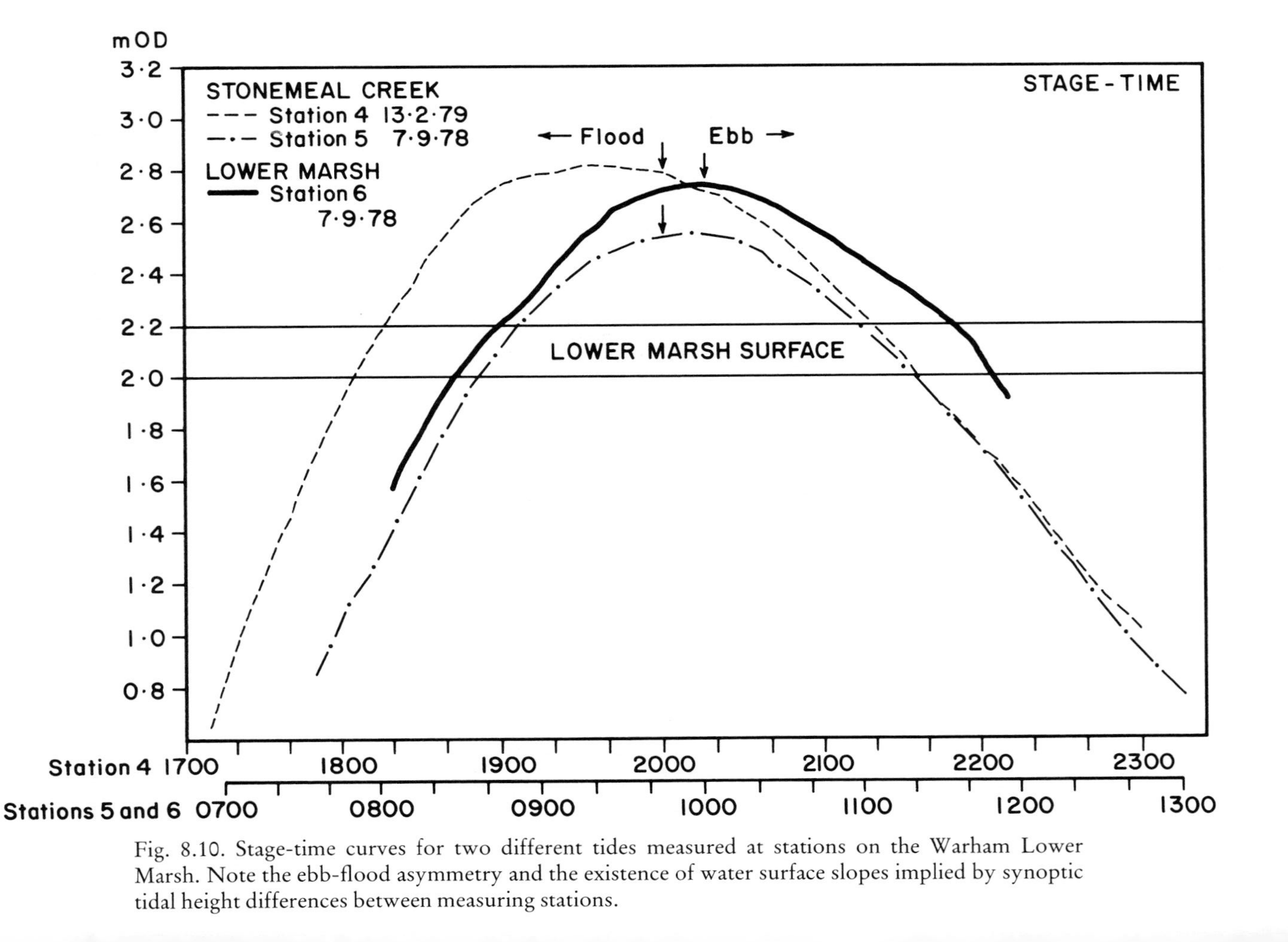

Fig. 8.10. Stage-time curves for two different tides measured at stations on the Warham Lower Marsh. Note the ebb-flood asymmetry and the existence of water surface slopes implied by synoptic tidal height differences between measuring stations.

The preferential deposition of sediment close to the main creeks leads to the formation of levees, which are preferentially colonised by *Halimione portulacoides*, along their banks, thereby creating enclosed depressions with relatively poor drainage. Parts of the most frequently waterlogged areas sometimes develop into circular or eliptical salt pans if the vegetation is killed off. Other salt pans, which usually have a more elongate and meandering form, originate through localised blocking of creeks due to bank failures (*cf.* Pethick, 1974).

Pethick (1969) suggested that the creek pattern developed on north Norfolk marshes after a period of 50–100 years remains relatively stable, although the cross-sectional profile of any given creek may continue to evolve. However, the very old (pre-Romano-British) marshes between Wells and Blakeney have a quite different drainage pattern to that on the younger marshes, being characterised by a relatively small number of highly sinuous creeks and numerous abandoned channels, some of which have evolved into elongate salt pans and others which have been infilled with sediment and colonised by vegetation (Fig. 8.9). At present it is uncertain whether the highly sinuous form of the creeks on the older marshes has been inherited from the early Holocene tidal flat and incipient marsh environments, or whether an initially dendritic network has gradually evolved into the present pattern as the marsh has grown vertically under conditions of rising sea level. However, observations of present-day processes and comparison of air photographs taken at different dates suggest that the sinuosity of certain creeks is still increasing due to bank failure on the outer bends of meanders and point bar accretion on the inside bends.

Current velocity and tidal stage measurements (Bayliss-Smith *et al.*, 1979; Healey *et al.*, 1981; Green *et al.*, 1986; French, 1989) have demonstrated that the tidal curves within the north Norfolk marshes display the asymmetry typical of many shallow water environments, in which the ebb tide lasts substantially longer than the flood tide (Fig. 8.10). Peak flood tidal flow velocitities within the marsh creeks (typically 0.5–0.6 m s^{-1} but exceptionally 0.8 m s^{-1}) occur close to the creek bed just as the tide rises above the bank full stage and spills onto the marsh surface. However, this velocity pulse is short lived and the near-bed creek velocity gradually falls to zero as the tide turns at high water (Fig. 8.11). The magnitude of the flood velocity pulse is greatest near the seaward end of the marsh drainage systems and decreases towards the landward margins (Healey *et al.*, 1981; French, 1989). Velocity pulses are not measurable on the

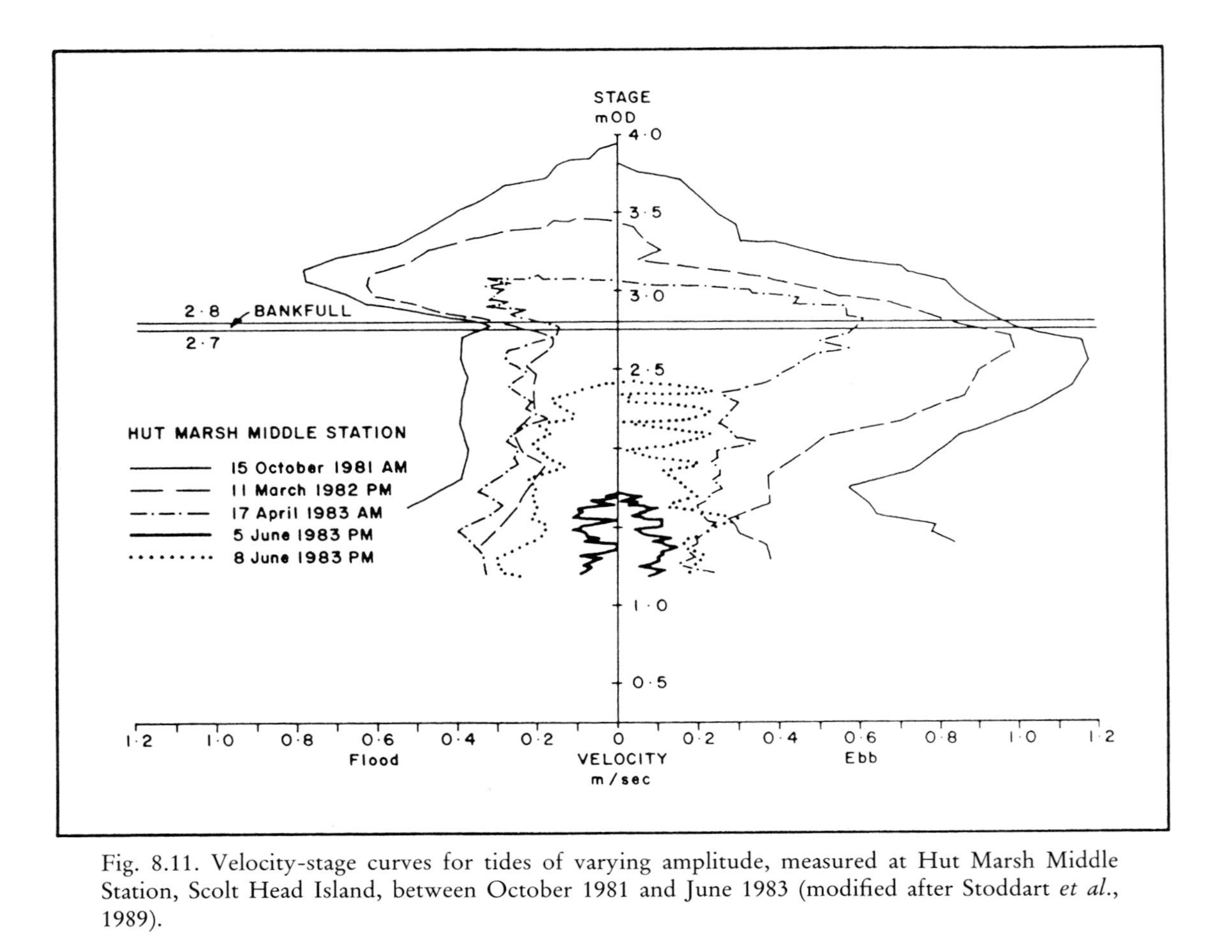

Fig. 8.11. Velocity-stage curves for tides of varying amplitude, measured at Hut Marsh Middle Station, Scolt Head Island, between October 1981 and June 1983 (modified after Stoddart *et al.*, 1989).

intertidal flats external to the marsh system. During overmarsh tides up to 40% of the water overlying the marsh at high tide may cross the marsh edge rather than entering through the creek system.

During spring tides, ebb tidal flow velocities within the creeks typically reach a maximum of up to 1.2 m s^{-1} when the water level has dropped below the marsh surface and the water surface slopes between the headwater creeks and the entrance to the marsh drainage system are at a maximum. Synoptic tidal height differences of 20 cm or more are commonly observed along individual creeks (Fig. 8.10). Channel scour by ebb currents is a major process which controls the depth of the creeks and their overall hydraulic geometry through its effect in initiating bank undercutting and collapse.

Relatively few data are available concerning suspended sediment concentrations in the north Norfolk marshes. Analysis of water samples from Hut Marsh on Scolt Head Island by French (1989) indicated maximum concentrations of about 1.0 g l^{-1} following a period of stormy conditions during winter. The principal factor controlling turbidity is wave action offshore and erosion of mud-bearing sediments on neighbouring sections of the coast, although higher concentrations also appear to be associated with spring tides, when tidal current velocities are higher, than with neap tides.

The mechanics of sediment deposition during overmarsh tides are still poorly understood, but probably involve a number of distinct processes whose relative importance is unknown. These include gravitational settling and flocculation during the period of slack water at high tide, downward diffusion to the sediment-water interface, physical interception by vegetation surfaces, and bio-deposition by filter-feeding organisms. The combined effect of these processes is to ensure that less sediment leaves the marsh on the ebb tide than enters on the flood. Preliminary estimates based on measurements based at Scolt Head Island (French, 1989) suggest that 20–80% of the sediment entering the marsh system on a typical spring tide may be deposited and remain within the marsh system, although there is much seasonal and even intertidal variation.

Measured short-term rates of sediment accretion on Scolt Head Island are of the order of 5–7 mm on Lower Hut Marsh and less than 2 mm a^{-1} on the somewhat higher Upper Hut Marsh (Steers, 1948; Stoddart *et al.*, 1989). Comparable rates of short-term vertical accretion in the range 3.7–5.3 mm a^{-1} were recorded on the Marams marshes at Blakeney by Carey and Oliver (1918) and Clymo (1967). Data collected by Steers (1948) for the period 1935–47 on Missel Marsh and

Hut Marsh indicated a reduction in average accretion rate with increasing marsh height. However, more recent data obtained from Hut Marsh for the period 1983–87 (Stoddart *et al.*, 1989) indicate no clear correlation between marsh height and accretion rate. These latter data suggest that, in the short term at least, large spatial variations in accretion rate are closely dependent on such factors as distance from major creeks which provide the major source of sediment.

Pethick (1980, 1981) examined a spatial sequence of marshes of differing ages between Blakeney and Scolt Head Island and demonstrated that vertical marsh growth is very rapid during the first 100 years after inception. Pethick concluded, however, that once the marsh surface elevation reaches a critical level in relation to the tidal frame, the rate of vertical marsh accretion slows dramatically. At this stage the marsh is considered to be 'mature'. According to Pethick, the marsh surface levels in north Norfolk tend towards an asymptote lying some 0.8 m below the level of the highest spring tides. However, work by Bayliss-Smith *et al.* (1979), Healey *et al.* (1981), Stoddart *et al.* (1989), French (1989) and the author (unpublished data) has shown that the marsh surface elevations given by Pethick are in error, some by as much as 0.75 m. The marsh heights and estimated marsh ages reported by Pethick (1981) and by French (1989, 1991) are shown in Table 8.1. Pethick (1981) also assumed a level tidal reference plane along the whole coast from Scolt to Blakeney. However, as pointed out previously in this paper, the height of predicted MHWS relative to Ordnance Datum decreases eastwards along the coast, being 3.75 m OD at Hunstanton, approximately 3.10 m OD at Brancaster Harbour, 3.0 m OD at Wells Bar, 2.65 m OD at Blakeney Bar and 2.45 m OD at Cromer. Respective heights for the predicted Highest Astronomical Tide are 3.85 m, 3.2 m, 3.1 m, 2.75 m and 2.55 m OD. Furthermore, there are significant differences in the height of high water (as measured under conditions undisturbed by a surge component) between the seaward and the landward margins of the marsh system. The maximum water level is often slightly higher near the seaward marsh edge, probably due to the greater frictional resistance offered to the incoming tide by the vegetated marsh surface and drainage system compared with the unvegetated tidal flats. In view of these spatial differences in the tidal reference level, it would seem inappropriate to apply an ergodic model, as used by Pethick, to investigate the temporal relationship between tidal frame and marsh elevation unless the elevations are expressed in proportional terms.

Table 8.1. *Mean surface height (m OD) and estimated age (years) of north Norfolk marshes. The age estimates, which are largely based on Ordnance Survey Six Inch Map and may not be entirely accurate, have been adjusted to a base year of 1991. The height estimates given by Pethick (1981) relate to* c. *1970 but underestimate the actual marsh heights at that time by 0.4–0.75 m. The height data reported by French (1989, 1991) relate to* c. *1981 and should also be regarded as minima.*

Marsh		Pethick (1981)		French (1989, 1991)	
		Age	Height	Age	Height
Blakeney	1	490 ± 50	2.43	—	—
Blakeney	2	81 ± 5	1.86	—	—
Blakeney	3	71 ± 1	1.72	—	—
Blakeney	4	141 ± 50	2.30	—	—
Stiffkey	5	>2000	2.46	>2000	2.85
Stiffkey	6	>2000	2.47	>2000	2.85
Stiffkey	7	41	1.18	39	1.90
Scolt Head Island					
Gt Aster	8	98 ± 13	1.75	94	2.68
Plantago	9	451 ± 120	2.30	—	—
Sueda	10	451 ± 120	2.26	—	—
Plover	11	451 ± 120	2.28	>462	2.96
Hut East	12	141 ± 30	2.05	—	—
Hut West	13	141 ± 30	2.12	—	—
Hut		—	—	96	2.78
Missel	14	98 ± 13	2.08	58	2.57
Spiral		—	—	94	3.01
N. Cockle Bight		—	—	33	2.39
Cockle Bight (tidal flat)		—	—	—	1.79

Comparison of the marsh elevation data in Table 8.1 with tidal heights given above demonstrates that most of the 'mature' marshes on the north Norfolk coast lie close to, or just above, the level of predicted MHWS. According to Admiralty predictions (Hydrographer of the Navy, 1991), approximately 3.5% of all high tides should exceed 2.6 m OD at Scolt Head Island (the approximate average level of Missel Marsh, Great Aster Marsh and Lower Hut Marsh), while at Wells and Warham approximately 3.5% of predicted high tides

Fig. 8.12. View over the eastern end of Holkham Meals and entrance to Wells Harbour in June 1984. Note the development of a new barrier system and incipient mud accumulation in its lee. Reproduced by permission of the Department of Air Photography, Cambridge University.

should exceed 2.35 m OD (the average level of the Warham Lower Marsh is approximately 2.2 m OD). In practice a slightly higher proportion of tides will exceed these levels due to frequent minor surge contributions to the resultant tides (data from the Wells tide gauge suggest that positive surge contributions are more frequent than negative surge contributions).

All of the younger north Norfolk marshes have formed on an intertidal sandflat substrate, often in the lee of a beach bar composed of sand or shingle. Changes along the shore between the entrance to Wells Harbour and Holkham Gap since the late 1970s demonstrate how rapid the initiation of a new marsh can be. Air photographs taken in 1984 (Fig. 8.12) show the early development of a dune-

Fig. 8.13. Incipient marsh forming on the mudflats behind the new barrier seaward of Holkham Meals (photograph taken in November 1990).

capped beach bar at the extreme eastern end of Holkham Meals. A thin (100 mm) drape of intertidal mud had accumulated in its lee at this time, but pioneer marsh vegetation was not established. By the autumn of 1990, however, the dunes on the incipient barrier had grown to a height of more than 3 m and a large blanket of mud, in places more than 250 mm thick, had been extensively colonised by *Salicornia* and *Aster* spp. (Fig. 8.13). Only time will show whether these features can withstand the effects of future winter storms and develop into a fully grown new marsh.

The nature of the vegetation zonation and succession on the Norfolk saltmarshes has been studied in detail by Chapman (1938, 1959) and Randerson (1975), and useful summaries are given in Chapman (1960), White (1989) and Woodell (1989). The intertidal flats are often partially colonised by eel grass (*Zostera*), seaweed (*Enteromorpha*) and several species of algae such as *Vaucheria* which render the surface less mobile and enable true pioneer marsh species to become established. The main natural pioneers on this coast include *Salicornia* spp. and *Suaeda maritima*, although stands of *Spartina anglica* also occur in places, including Morston and Blakeney. These are followed at slightly higher levels by *Aster tripolium*, *Halimione portulacoides*, *Limonium* spp., and *Salicornia perennis*. At still higher levels the main species are *Puccinellia maritima*, *Festuca* spp., *Triglochin* spp., *Spergularia* spp., *Armeria maritima*, and *Juncus* spp. These comunities in turn give way at the margins of the marshes to *Suaeda fruticosa* and *Agropyron* spp. on well-drained sites and to *Phragmites* and *Scirpus* communities at the transition to freshwater marshes.

Coastal management problems

As in many other parts of the United Kingdom, environmental management of the north Norfolk coast involves a need to reconcile the often conflicting objectives posed by conservationists, private individuals and companies with commercial interests, engineers faced with statutory obligations for coastal protection and flood defence, and the general public who wish to use the coast for an increasing variety of recreational purposes (Bayliss-Smith, 1990). The area is one of outstanding natural beauty which, so far, has escaped the worst effects associated with recreational pressures which have seriously affected certain other parts of the country. As such, the diverse, dynamic and relatively unspoiled character of its habitats make the area one of great ecological and scientific importance. However, recreational pressures are likely to increase in future years, and there is increasing concern that the effects of global warming may have significant consequences for both the wildlife and human communities along the coast. Mean sea level in north Norfolk is presently rising by approximately 1.5 mm a^{-1}, but this rate may increase by a factor of two or more during the next century if global warming and thermal expansion of the oceans become a reality. In addition, changes in wind and wave climate could also have a major effect on the sediment transport regime, lead to an increase in the frequency and magnitude of storm surge events, and cause long-term changes in the pattern of erosion and sedimentation.

For more than two thousand years, coastal accretion in The Wash and north Norfolk has been fed largely by erosion of coastal cliffs in Lincolnshire, Humberside and northeast Norfolk. The great success of land reclamation schemes in The Wash and north Norfolk has been accomplished largely due to a continued natural sediment supply from areas to the north and east. Extensive coastal protection works have been undertaken in these sediment source areas since the 1953 floods, and the prospect of further works remains. There is already

Fig. 8.14. The breach in the sea wall at Burnham Overy Staithe during the 1953 storm surge which caused extensive flooding to the reclaimed marshes as far east as Holkham (reproduced from Pollard, 1978).

evidence that low water mark is moving landwards in many places on the East Anglian coast, accompanied by an increase in the gradient of the nearshore profile (Anglian Water, 1988a, 1988b). This is a trend which can only be ehanced by the reduction in sediment supply due to shore protection works (Clayton, 1980, 1989b, 1990; Wakelin, 1989; Fleming, 1989). In north Norfolk, sections of the depositional coast near Gore Point, Titchwell, Brancaster and Stiffkey are experiencing erosion, although accretion is occurring elsewhere, notably between Wells and Burham Overy Staithe. It seems probable that shoreline retreat will spread, and possibly accelerate, in the next 100 years as the effects of reduced sediment supply become more apparent and as sea level starts to rise at a more rapid rate. Under these conditions, the higher, more landward marshes adjoining the sea defences could continue to grow vertically at a rate equivalent to the rate of sea-level rise if they receive sufficient fine sediment eroded from the more seaward marshes and mudflats.

All low-lying parts of the East Anglian coast, including north Norfolk, are prone to flooding during severe storm surges as occurred on the night of 31 January/1 February 1953 (Steers, 1953; Pollard, 1978; Summers, 1978). On this occasion the tide reached a level of 5.13 m OD at Wells Quay, 5.18–5.49 m OD at Holme

Fig. 8.15. The washover apron formed near the eastern end of Scolt Head Island during the 1978 storm surge. A similar breakthrough occurred during the 1953 surge (reproduced by permission of the Cambridge University Committee for Air Photography).

marshes, 5.49 m OD at Burnham Overy Staithe (Jensen, 1953) and 4.88 m OD at Blakeney (Grove, 1953). Even higher levels, up to 5.51 m OD, were recorded locally during the January 1978 storm surge (Steers *et al.*, 1979). Although the sea walls surrounding the reclaimed marshes, whose crests mostly stand at an elevation of between 4.5 and 5.3 m OD, were not directly overtopped, several were breached, leading to flooding of the marshland behind (Fig. 8.14). Other significant floods of varying severity occurred in 1817, 1883, 1897, 1912, 1938, 1949, 1961, 1962, 1969, and 1976. Following the 1953 and 1978 storm surges, many of the man-made sea defences have been strengthened, but the risk of flooding remains around the harbours at Wells and Blakeney. Since the saltmarshes act as a protective buffer which absorbs tide and wave energy, a significant narrowing or lowering of the marshes (relative to the moving tidal frame) would be likely to increase the storm surge levels and wave action at the landward limit of the marshes. Marsh surface levels and widths should therefore be monitored closely over the next century.

If the rate of sea-level rise increases significantly, or if there is a significant increase in the frequency of major storm events, landward migration of the major barrier features, including Blakeney spit and Scolt Head Island, is also likely to accelerate. Storm breaching and washover of the barriers, as occurred in 1953 and 1978 (Fig. 8.15), will become more frequent and many become permanent. This will result in significant ecological changes, particularly on the Cley, Blakeney and eastern Scolt marshes, and may also increase the flooding risk to property along the landward fringe of the marshes. However, in view of the relatively small number of properties and low agricultural value of the land involved, large-scale sea defence works appear to be unjustified for the forseeable future. Accelerated erosion on some parts of the coast is also likely to be offset by continued accretion elsewhere. For example, coastal progradation between Wells and Burnham Overy Staithe has been substantial during the past 20 years, and may well continue in the face of accelerated sea-level rise if a postive local sediment budget can be sustained.

Acknowledgements

Research on the north Norfolk coast has been supported by the UK Natural Environment Research Council (Grants GR3/6016 and GR9/296), the Royal Society, and the Universities of Cambridge and Reading. I thank the National Rivers Authority (Anglia Region), Proudman Oceanographic Laboratory, Birkenhead, and Sir William Halcrow and Partners, Swindon, for provision of tidal and unpublished survey data. Dr J. French kindly gave permission to make reference to his partly unpublished data relating to Scolt Head Island. Permission to work on the coast was granted by the Nature Conservancy Council and the National Trust. This paper represents University of Reading PRIS Contribution No. 142.

References

ADMIRALTY (1988) *Admiralty Chart 108. Skegness to Blakeney including the Wash. Admiralty Hydrographic Department.*

ALLISON, H. (1985) *The Holocene Evolution of Scolt Head Island.* Unpublished PhD Thesis, University of Cambridge.

ALLISON, H. (1989) The sedimentary history of Scolt Head Island. In H. Allison & J.P. Morley (eds.) *Blakeney Point and Scolt Head Island.* The National Trust, Blickling, Norfolk, 28–32.

ANGLIAN WATER (1988a) *Anglian Coastal Management Atlas.* Sir William Halcrow and Partners, Swindon.

ANGLIAN WATER (1988b) *The Sea Defence Management Study for the Anglian Region. Supplementary Studies Report.* Sir William Halcrow and Partners, Swindon.

BARFOOT, P.J. & TUCKER, J.J. (1980) Geomorphological changes at Blakeney Point, Norfolk. *Transactions of the Norfolk and Norwich Naturalists Society* **25**, 49–60.

BARNES, R.S.K., SATELLE, D.B., EVERTON, I.J., NICHOLAS, W. & SCOTT, D.H. (1976) Intertidal sands and interstitial fauna associated with different stages of salt marsh development. *Estuarine, Coastal and Marine Science* **4**, 497–512.

BAYLISS-SMITH, T.P. (1990) Coastal change. In T.P. Bayliss-Smith & S. Owens (eds) *Britain's Changing Environment from the Air.* Cambridge University Press, Cambridge, 77–99.

BAYLISS-SMITH, T.P., HEALEY, R.G., LAILEY, R., SPENCER, T. & STODDART, D.R. (1979) Tidal flows in salt marsh creeks. *Estuarine, Coastal and Marine Science* **9**, 235–255.

CAMBERS, G. (1976) Temporal scales in coastal erosion systems. *Transactions of the Institue of British Geographers* N.S. **1**, 246–256.

CAREY, A.E. & OLIVER, F.W. (1918) Blakeney Point, Norfolk, from an engineering point of view. In A.E. Carey & F.W. Oliver *Tidal Lands: A Study of Shore Problems.* Blackie, Glasgow, 218–241.

CHAPMAN, V.J. (1938) Studies in saltmarsh ecology. Sections I-III. *Journal of Ecology* **26**, 144–179.

CHAPMAN, V.J. (1959) Studies in saltmarsh ecology. IX. Changes in saltmarsh vegetation at Scolt Head Island, Norfolk. *Journal of Ecology* **47**, 619–639.

CHAPMAN, V.J. (1960) *Saltmarshes and Salt Deserts of the World.* Leonard Hill, London, 392 pp.

CLAYTON, K. (1976) Norfolk sandy beaches, applied geomorphology and the engineer. *Bulletin of the Geological Society of Norfolk* **28**, 49–67.

CLAYTON, K. (1980) Coastal protection along the East Anglian coast, U.K. *Zeitschrift für Geomorphologie Supplement Band* **34**, 165–172.

CLAYTON, K. (1989a) Sediment input from the Norfolk cliffs, eastern

England – a century of protection and its effect. *Journal of Coastal Research* **5**, 433–442.

CLAYTON, K. (1989b) The implications of climatic change. In *Coastal Management.* Thomas Telford, London, 165–176.

CLAYTON, K. (1990) Sea level rise and coastal defences in the U.K. *Quarterly Journal of Engineering Geology* **23**, 283–287.

CLYMO, R.S. (1967) Accretion rate in two of the saltmarshes at Blakeney Point, Norfolk. *Transactions of the Norfolk and Norwich Naturalists Society* **21**, 3–6.

COZENS-HARDY, B. (1924–5) Cley-next-the-Sea and its marshes. *Transactions of the Norfolk and Norwich Naturalists Society* **12**, 354–373.

COZENS-HARDY, B. (1972) Havens in North Norfolk. *Norfolk Archaeology* **35**, 356–363.

FLEMING, C.A. (1989) The Anglian Sea Defence Management Study. In *Coastal Management.* Thomas Telford, London, 153–164.

FOSTER, W.A. (1989) Terrestrial animals of Blakeney Point and Scolt Head Island. In H. Allison & J.P. Morley (eds.), *Blakeney Point and Scolt Head Island.* The National Trust, Blickling, Norfolk, 76–86.

FRENCH, J.R. (1989) *Hydrodynamics and Sedimentation in a Macro-tidal Salt Marsh, North Norfolk.* Unpublished PhD Thesis, University of Cambridge.

FRENCH, J.R. (1991) Eustatic and neotectonic controls on saltmarsh sedimentation. In N.C. Kraus (ed.) *Coastal Sediments* '91. American Society of Civil Engineers (in press).

FRENCH, J.R., SPENCER, T. & STODDART, D.R. (1990) Backbarrier marshes of the north Norfolk coast: geomorphic development and response to rising sea levels. *University College London, Discussion Papers in Conservation* 54, 28 pp.

FRID, C.L.J. (1988) The marine fauna of the north Norfolk saltmarshes and their ecology. *Transactions of the Norfolk and Norwich Naturalists Society* **28**, 46–50.

FUNNELL, B.M. & PEARSON, I. (1984) A guide to the Holocene geology of north Norfolk. *Bulletin of the Geological Society of Norfolk* **34**, 123–140.

FUNNELL, B.M. & PEARSON, I. (1989) Holocene sedimentation on the north Norfolk barrier coast in relation to relative sea level changes. *Journal of Quaternary Science* **4**, 25–36.

GALE, S.J., HOARE, P.G., HUNT, C.O. & PYE, K. (1988) The Middle and Upper Quaternary deposits at Morston, north Norfolk. *Geological Magazine* **125**, 521–533.

GALLOIS, R.W. (1976) The Pleistocene history of west Norfolk. *Bulletin of the Geological Society of Norfolk* **30**, 3–38.

GODWIN, H. (1934) Discussion of J.A. Steers' paper on Scolt Head Island. *Geographical Journal* **83**, 494–498.

GODWIN, H. (1943) Coastal peat beds of the British Isles and North Sea. *Journal of Ecology* **31**, 199–247.

GREEN, H.M., STODDART, D.R., REED, D.J. & BAYLISS-SMITH, T.P. (1986) Salt marsh tidal creek dynamics, Scolt Head Island, Norfolk, England. In G. Sigbjarnarson (ed.), *Iceland Coastal and River Symposium Proceedings.* National Energy Authority, Rejkjavik, 93–103.

GROVE, A.T. (1953) The sea flood on the coasts of Norfolk and Suffolk. *Geography* **38**, 164–170.

HALES, J. & SIMMS, N. (1990) *Blakeney Point and the Glaven Ports.* Orlando Publishing, Briston, Norfolk, 14 pp.

HARDY, J.R. (1964) The movement of beach material and wave action near Blakeney Point, Norfolk. *Transactions of the Institute of British Geographers* **34**, 53–70.

HEALEY, R.G., PYE, K., STODDART, D.R. & BAYLISS-SMITH, T.P. (1981) Velocity variations in salt marsh creeks, Norfolk, England. *Estuarine, Coastal and Shelf Science* **13**, 535–545.

HUNT, O.D. (1971) Holkham Salts Hole, an isolated salt-water pond with relict features. An account of studies by the late C.F.A. Pantin. *Journal of the Marine Biological Association of the United Kingdom* **51**, 717–741.

HYDROGRAPHER OF THE NAVY (1991) *Admiralty Tide Tables 1991.* Admiralty, London.

JEFFERIES, R.L. (1977) Growth response of coastal halophytes to organic carbon. *Journal of Ecology* **65**, 847–865.

JEFFERIES, R.L. & PERKINS, N. (1977) The effects on the vegetation of the additions of inorganic nutrients to salt marsh soils at Stiffkey, Norfolk. *Journal of Ecology* **65**, 867–883.

JENSEN, H.A.P. (1953) Tidal inundations past and present. *Weather* **8**, 85–89, 108–113.

MCCAVE, I.N. (1977) *Sediments of the East Anglian Coast.* East Anglian Coastal Research Program Report No. 6. School of Environmental Sciences, University of East Anglia, Norwich.

MCCAVE, I.N. (1978) Grain-size trends and transport along beaches: example from eastern England. *Marine Geology* **28**, M43–M51.

MCCAVE, I.N. (1987) Fine sediment sources and sinks around the East Anglian coast. *Journal of the Geological Society* **144**, 149–152.

OLIVER, F.W. (1923) Report of the Blakeney Point Research Station, for the years 1920–23. *Transactions of the Norfolk and Norwich Naturalists Society* **17**, 206–216.

OLIVER, F.W. & SALISBURY, E.J. (1913) The topography and vegetation of the National Trust Reserve known as Blakeney Point, Norfolk. *Transactions of the Norfolk and Norwich Naturalists Society* **9**, 485–542.

PEARSON, I., FUNNELL, B.M. & MCCAVE, I.N. (1989) Sedimentary environments of the sandy barrier/tidal marsh coastline of north Norfolk. *Bulletin of the Geological Society of Norfolk* **39**, 3–44.

PETHICK, J.S. (1969) Drainage in tidal marshes. In J.A. Steers, *The*

Coastline of England and Wales, 2nd edn. Cambridge University Press, Cambridge, 725–730.

PETHICK, J.S. (1974) The distribution of salt pans on tidal marshes. *Journal of Biogeography* **1**, 57–62.

PETHICK, J.S. (1980) Salt marsh initiation during the Holocene transgression: the example of the north Norfolk marshes, England. *Journal of Biogeography* **7**, 1–9.

PETHICK, J.S. (1981) Long term accretion rates on tidal marshes. *Journal of Sedimentary Petrology* **51**, 571–577.

POLLARD, M. (1978) *North Sea Surge. The Story of the East Coast Floods of 1953.* Dalton, Lavenham, Suffolk, 136 pp.

PURCHAS, A.W. (1965) *Some History of Wells-next-the-Sea and District.* East Anglian Magazine Ltd., Ipswich, 140 pp.

PYE, K. (1981) Marshrock formed by iron sulphide and siderite cementation in saltmarsh sediments. *Nature* **294**, 650–652.

PYE, K. (1984) SEM analysis of siderite cements in intertidal marsh sediments, Norfolk, England. *Marine Geology* **56**, 1–12.

PYE, K. (1988) An occurrence of akaganeite (B-FeO.OH.Cl) in Recent oxidized carbonate concretions, Norfolk, England. *Mineralogical Magazine* **52**, 125–126.

PYE, K. (1991) *A Scientific Bibliography of the North Norfolk Coast.* Cambridge Environmental Research Consultants Ltd. (Publications), Cambridge, 14 pp.

PYE, K., DICKSON, J.A.D., SCHIAVON, N., COLEMAN, M.L. & COX, M. (1990) Formation of siderite-Mg-calcite-iron monosulphide concretions in intertidal marsh and sandflat sediments, north Norfolk, England. *Sedimentology* **37**, 325–343.

RANDERSON, P.F. (1975) *An Ecological Model of Succession on a Norfolk Salt Marsh.* Unpublished PhD Thesis, University of London.

ROY, P.S. (1967) *The Recent Sedimentology of Scolt Head Island, Norfolk.* Unpublished PhD Thesis, University of London.

SEAGO, M.J. (1989) Birds of Blakeney Point and Scolt Head Island. In H. Allison & J.P. Morley (eds) *Blakeney Point and Scolt Head Island.* The National Trust, Blickling, Norfolk, 87–108.

SHAW, H.F. (1973) Clay mineralogy of Quaternary sediments in the Wash embayment, eastern England. *Marine Geology* **14**, 29–45.

SHENNAN, I. (1989) Holocene crustal movements and sea level changes in Great Britain. *Journal of Quaternary Science* **4**, 77–89.

SOLOMON, J.D. (1931) Palaeolithic and Mesolithic sites at Morston, north Norfolk. *Man* **31**, 275–278.

STEERS, J.A. (1927) The East Anglian coast. *Geographical Journal* **69**, 24–43.

STEERS, J.A. (1929) Geographical work on Scolt Head Island and adjacent areas. *Transactions of the Norfolk and Norwich Naturalists Society* **12**, 664–667.

STEERS, J.A. (ed.) (1934a) *Scolt Head Island.* Heffer, Cambridge, 234 pp.

STEERS, J.A. (1934b) Scolt Head Island. *Geographical Journal* **83**, 479–494.

STEERS, J.A. (1935) A note on the rate of sedimentation on a salt marsh on Scolt Head Island, Norfolk. *Geological Magazine* **72**, 443–435.

STEERS, J.A. (1936) Some notes on the north Norfolk coast from Hunstanton to Brancaster. *Geographical Journal* **87**, 35–46.

STEERS, J.A. (1948) Twelve years measurements of accretion on Norfolk saltmarshes. *Geological Magazine* **85**, 163–166.

STEERS, J.A. (1953) The east coast floods, January 31 – February 1, 1953. *Geographical Journal* **119**, 280–298.

STEERS, J.A. (1969) Hunstanton to Reculver. In J.A. Steers, *The Coastline of England and Wales*, 2nd edn. Cambridge University Press, Cambridge, 345–405.

STEERS, J.A. (1989) The physical features of Scolt Head Island and Blakeney Point. In H. Allison & J.P. Morley (eds) *Blakeney Point and Scolt Head Island.* The National Trust, Blickling, Norfolk, 14–27.

STEERS, J.A., STODDART, D.R., BAYLISS-SMITH, T.P., SPENCER, T. & DURBIDGE, P.M. (1979) The storm surge of 11 January 1978 on the east coast of England. *Geographical Journal* **145**, 192–205.

STODDART, D.R., REED, D.J. & FRENCH, J.R. (1989) Understanding salt-marsh accretion, Scolt Head Island, Norfolk, England. *Estuaries* **12**, 228–236.

STRAW, A. (1960) Limit of the last glaciation in Norfolk. *Proceedings of the Geologists Association* **71**, 379–390.

SUMMERS, D. (1978) *The East Coast Floods.* David and Charles, Newton Abbott.

TREHERNE, J.E. & FOSTER, W.A. (1979) Adaptive strategies of air-breathing arthropods from marine saltmarshes. In R.L. Jefferies (ed.) *Ecological Processes in Coastal Environments.* Blackwell, Oxford, 165–173.

VINCENT, C.E. (1979) Longshore sand transport rates – a simple model for the East Anglian coastline. *Coastal Engineering* **3**, 113–136.

WHITE, D.J.B. (1989) The botany and plant ecology of Blakeney Point. In H.A. Allison & J.P. Morley (eds) *Blakeney Point and Scolt Head Island.* The National Trust, Blickling, Norfolk, 33–48.

WILLIAMS, R.B. & BEALE, C.J. (1976) A note on Holkham Lake, Norfolk. *Transactions of the Norfolk and Norwich Naturalists Society* **23**, 83.

WOODELL, S.R.J. (1989) Vegetation and plants of Scolt Head Island. In H. Allison & J.P. Morley (eds) *Blakeney Point and Scolt Head Island.* The National Trust, Blickling, Norfolk, 49–58.

WOODWARD, H.B. (1884) *The Geology of the Country Around Fakenham, Wells and Holt.* Memoir of the Geological Survey of England and Wales. HMSO, London.

Appendix 1

Saltmarsh workshop – Participants

ALLEN, PROFESSOR J.R.L., Postgraduate Research Institute for Sedimentology, The University, PO Box 227, Whiteknights, Reading RG6 2AB.

ANDREWS, DR J.E., School of Environmental Sciences, University of East Anglia, Norwich NR4 7TJ.

ARTHURTON, MR R.S., Coastal Geology Group, British Geological Survey, Keyworth, Nottingham NG12 5GG

BLACK, MR K., School of Ocean Sciences, MENAI BRIDGE, Anglesey, Gwynedd LL59 5EY.

BONNETT, DR P., Environment and Energy Division, B364 Harwell Laboratory, Oxon OX11 0RA.

BOORMAN, DR L.A., Institute of Terrestrial Ecology, Monks Wood Experimental Station, Huntingdon, Cambs PE17 2LS.

BRAMPTON, DR A.H., Coastal Engineering Group, Hydraulics Research Ltd, Wallingford, Oxon OX10 8BA.

BRAY, MR M., Department of Geography, Buckingham Building, Portsmouth Polytechnic, Lion Terrace, Portsmouth PP1 3HE.

BURD, MS F., English Nature, Northminster House, Peterborough PE2 0JY.

CARPENTER, MS K., School of Environmental Sciences, University of East Anglia, Norwich NR4 7TJ.

DIXON, MR A.M., National Rivers Authority, Threshelfords Business Park, Inworth Road, Feering, Colchester CO5 9SE.

DOODY, DR P., Coastal Ecology Branch, Chief Scientist Directorate, English Nature, Northminster House, Peterborough PE1 1UA.

DUFFY, MR M., Postgraduate Research Institute for Sedimentology, The University, PO Box 227, Whiteknights, Reading RG6 2AB.

DYER, PROFESSOR K.R., Institute of Marine Studies, Polytechnic South West, Plymouth PL4 8AA.

EVANS, DR C.D.R., Coastal Geology Group, British Geological Survey, Keyworth, Nottingham NG12 5GG.

FLETCHER, MS C., Department of Civil Engineering, Imperial College Road, London SW7 2BU.

FREEMAN, MR D., School of Civil Engineering, University of Birmingham, Birmingham B15 2TT.

FRITH, MR C.W., Binnie & Partners, 69 London Road, Redhill, Surrey RH1 1LQ.

FUNNELL, PROFESSOR B.M., School of Environmental Sciences, University of East Anglia, Norwich NR4 7TJ.

GRANT, DR A., School of Environmental Sciences, University of East Anglia, Norwich NR4 7TJ.

GRAY, DR A.J., Institute of Terrestrial Ecology, Furzebrook Research Station, Furzebrook Road, Wareham, Dorset BH20 5AS.

HICKLING, MR J., English Nature, Blackwell, Bowness-on-Windermere, Cumbria LA23 3JR.

HILL, MR C., Geodata Institute, University of Southampton, Southampton SO9 5NH.

HUTCHINSON, DR S.M., Environmental Science Division, University of Lancaster, Lancaster LA1 4YQ.

JICKELLS, DR T., School of Environmental Sciences, University of East Anglia, Norwich NR4 7TJ.

KE XIANKUN, MR, Department of Oceanography, The University of Southampton, Southampton SO9 5NH.

KIRBY, DR R., Ravensrodd Consultants Ltd, 6 Queens Drive, Taunton, Somerset TA1 4XW.

LAMBERT, MR R., 77 River View, Tarleton, Preston PR4 6ED.

LEGGETT, MR D.J., National Rivers Authority, Kingfisher House, Goldelay Way, Orton, Goldelay, Peterborough PE2 0ZR.

MACGUIRE, MS F., School of Environmental Sciences, University of East Anglia, Norwich NR4 7TJ.

MCKEW, MR J., Sir William Halcrow and Partners Ltd, Burderop Park, Swindon, Wiltshire SN4 0QD.

MCNAUGHT, MR K., English Nature, North West Regional Office, Blackwell, Bowness-on-Windermere, Cumbria LA23 3JR.

MUIRHEAD, DR S.J., Energy Technology Support Unit, B156 Harwell Laboratory, Oxon OX11 0RA.

NEAL, MR A., Postgraduate Research Institute for Sedimentology, The University, PO Box 227, Whiteknights, Reading RG6 2AB.

OTTO, MR S., Postgraduate Research Institute for Sedimentology, The University, PO Box 227, Whiteknights, Reading RG6 2AB.

PETHICK, DR J., Institute of Estuarine & Coastal Studies, University of Hull, Hull HU6 7RX.

PRINGLE, DR A.W., Department of Geography, Lancaster University, Lancaster LA1 4YB.

PYE, DR K., Postgraduate Research Institute for Sedimentology, The University, PO Box 227, Whiteknights, Reading RG6 2AB.

RADLEY, DR G., English Nature, Northminster House, Peterbroough PE2 0JY.

RUDDY, MR G., School of Environmental Sciences, University of East Anglia, Norwich NR4 7TJ.

SMITH, MR A.J.E., Binnie & Partners, 69 London Road, Redhill, Surrey RH1 1LQ.

SPENCER, DR T., Department of Geography, Cambridge University, Downing Place, Cambridge CB2 3EN.

STEVENS, MR N.W., National Rivers Authority, Wessex Region, Rivers House, East Quay, Bridgwater, Somerset.

THOMAS, DR E., Department of Earth Sciences, University of Cambridge, Downing Street, Cambridge CB2 3EQ.

TOFT, MRS A., Sir William Halcrow and Partners Ltd, Burderop Park, Swindon, Wiltshire SN4 0QD.

TOOLEY, DR M.J., Department of Geography, University of Durham, Science Laboratories, South Road, Durham DH1 3LE.

TUBBS, MR C.R., English Nature, 1 Southampton Road, Lyndhurst, Hampshire SO43 7BU.

WATKINS, MR M.G., County Planning Department, Hampshire County Council, The Castle, Winchester, Hampshire SO23 8UE.

WHEELER, MR A.J., Department of Earth Sciences, The University, Cambridge CB2 3EQ.

YASIN, MR A.R., Postgraduate Research Institute for Sedimentology, The University, PO Box 227, Whiteknights, Reading RG6 2AB.

ZHONG SHI, MR, Institute of Earth Studies, University College of Wales, Aberystwyth, Dyfed SY23 3DB.

Appendix 2

University of Reading
Postgraduate Research Institute for Sedimentology

Saltmarsh workshop – Posters

DR L.A. BOORMAN (Monks Wood Experimental Station) *Comparative study of saltmarsh processes*

MS K. CARPENTER & DR T. JICKELLS (University of East Anglia) *A method to measure nutrient flux, as used on a Norfolk saltmarsh*

MR C.W. FRITH & MR A.J.E. SMITH (Binnie & Partners) *Binnie & Partners*

MR S. OTTO (University of Reading) *Erosion of saltmarshes along the Severn Estuary, SW Britain*

DR A.W. PRINGLE (University of Lancaster) *Saltmarsh accretion and erosion in Morecambe Bay, NW England*

DR E. THOMAS (University of Cambridge), DR J.C. VAREKAMP & DR O. VAN DER PLASSCHE *Sea-level rise and climate fluctuations over the last 2000 years: records from marshes*

MR A.R. YASIN (University of Reading) *A sedimentologic study of saltmarshes in a sand-dominated estuarine system: the Duddon Estuary, NW England*

Index